RAND PROJECT AIR FORCE

# DO JOINT FIGHTER PROGRAMS SAVE MONEY?

**Mark A. Lorell, Michael Kennedy, Robert S. Leonard, Ken Munson, Shmuel Abramzon, David L. An, Robert A. Guffey**

Prepared for the United States Air Force

Approved for public release; distribution unlimited

The research described in this report was sponsored by the United States Air Force under Contract FA7014-06-C-0001. Further information may be obtained from the Strategic Planning Division, Directorate of Plans, Hq USAF.

**Library of Congress Cataloging-in-Publication Data** is available for this publication.

ISBN: 978-0-8330-7456-0



*Cover image courtesy of iStockphoto.*



RAND OFFICES

SANTA MONICA, CA • WASHINGTON, DC

PITTSBURGH, PA • NEW ORLEANS, LA • JACKSON, MS • BOSTON, MA

DOHA, QA • CAMBRIDGE, UK • BRUSSELS, BE

www.rand.org

# Preface

In the past 50 years, the U.S. Department of Defense (DoD) has pursued numerous joint aircraft programs, the largest and most recent of which is the F-35 Joint Strike Fighter (JSF). Joint aircraft programs, in which two or more services participate in the development, procurement, and sustainment of a single aircraft design, are thought to save significant Life Cycle Cost (LCC) by eliminating duplicate efforts and realizing economies of scale.

But the need to accommodate different service requirements in a single design or common design family leads to greater program complexity, increased technical risk, and common functionality or increased weight beyond that needed for some variants, potentially leading to higher overall cost, despite the efficiencies. The fundamental question we seek to answer is how much savings accrue, on average, from joint aircraft programs and are they sufficient to offset the additional costs arising from greater complexity? In short, do joint fighter and other aircraft programs cost less overall throughout their entire life cycles than an equivalent set of specialized, single-service systems?

To help Air Force leaders (and acquisition decisionmakers in general) select an appropriate acquisition strategy for future combat aircraft, Gen Donald Hoffman, as former commander of Air Force Materiel Command (AFMC), asked RAND Project AIR FORCE to analyze the costs and savings of joint tactical aviation acquisition programs. The study team examined whether historical joint aircraft programs, and JSF in particular, have saved LCC compared with comparable single-service programs. Also examined were the implications of joint fighter programs for the health of the industrial base and for operational and strategic risk. Greater detail on the methodologies and approaches used in the analysis reported in this document has been published in a separate document.[1]

Our analysis shows that historical joint aircraft programs have not saved overall LCC versus single-service aircraft programs and that JSF is currently not on the path to saving LCC versus comparable notional single-service fighter programs. The objective of this research was to inform the formulation of acquisition strategies and program structures for future fighter and other aircraft major defense acquisition programs. These findings do not, therefore, directly address policy questions regarding the current or future status of JSF. However, they can provide insight into how and why JSF has arrived at its current status with respect to cost and cost growth, and they could contribute to broader analysis seeking to develop policy options for the way ahead for the JSF program.

The research reported here was sponsored by Gen Donald Hoffman, former Commander of Air Force Materiel Command, and conducted within the Resource Management Program of RAND Project AIR FORCE as part of the "Cost/Benefit Analysis of Joint Tactical Aviation Acquisition Programs" project.

This document should be of interest to the senior defense acquisition and cost analysis policy communities.

## RAND Project AIR FORCE

RAND Project AIR FORCE (PAF), a division of the RAND Corporation, is the U.S. Air Force's federally funded research and development center for studies and analyses. PAF provides the Air Force with independent analyses of policy alternatives affecting the development, employment, combat readiness, and support of current and future air,

[1] See Mark A. Lorell, Michael Kennedy, Robert S. Leonard, Ken Munson, Shmuel Abramzon, David L. An, and Robert A. Guffey, *Do Joint Fighter Programs Save Money? Technical Appendixes on Methodology*, Santa Monica, Calif.: RAND Corporation, MG-1225/1-AF, 2013.

space, and cyber forces. Research is conducted in four programs: Force Modernization and Employment; Manpower, Personnel, and Training; Resource Management; and Strategy and Doctrine.

Additional information about PAF is available on our website: http://www.rand.org/paf/

# Contents

# Figures

# Tables

# Summary

The U.S. Department of Defense (DoD) has launched or attempted to launch numerous joint fighter and other joint aircraft programs in the past 50 years.[1] The largest and most recent is the F-35 Joint Strike Fighter (JSF), which was designed for use by the U.S. Air Force, U.S. Navy, U.S. Marine Corps, and international partners and is currently in low-rate initial production. The main purpose of a joint program, versus a set of single-service programs, is to save overall Life Cycle Cost (LCC) by eliminating duplicate research, development, test, and evaluation (RDT&E) efforts and achieving economies of scale in procurement and operations and support (O&S). Yet, the need to integrate multiple service requirements in a single design increases the complexity of joint programs and potentially leads to higher-than-average cost growth that could reduce or even negate potential savings.

There have been no comprehensive assessments of costs and savings in historical joint aircraft programs based on actual joint aircraft program outcomes and historical cost data. To help inform future acquisition strategies for fighter aircraft, the commander of Air Force Materiel Command (AFMC) asked RAND Project AIR FORCE (PAF) to analyze the costs and benefits of historical joint aircraft programs, from the early 1960s through today's JSF. The project addressed five major questions:

---

1 Our definition of a fully joint aircraft program is an aircraft program in which two or more services are significantly involved in all stages of the acquisition and operational phases—that is, in the design, development, procurement, and operations and support (O&S) of the aircraft.

- Have historical joint aircraft programs saved LCC compared with comparable single-service aircraft programs?
- Is JSF on track to save LCC compared with notional equivalent single-service fighter programs?
- What factors contributed to cost outcomes in historical joint aircraft programs, as well as JSF?
- What are the implications of a joint aircraft approach for the industrial base?
- What are the implications of a joint aircraft approach for operational and strategic risk?[2]

The project performed quantitative and qualitative reviews of joint fighter and other joint aircraft programs, drawing from a broad range of published sources, as well as the RAND Selected Acquisition Report (SAR) database (SARs document cost and schedule estimates at different stages; the RAND SAR database includes more than 300 major defense acquisition programs) and databases of historical O&S costs for Air Force aircraft. The data-cutoff point for this analysis was November 2011. Greater detail on the methodologies and approaches used in the analysis reported in this document has been published in a separate document.[3]

Because of the impossibility of making direct cost comparisons, in which multiple similar single-service programs were developed in parallel with an equivalent joint program, PAF sought to answer the question of which approach costs less by comparing the cost growth of joint and single-service programs. If cost growth tends to differ and be higher for joint programs, this would suggest that the difficulties

---

[2] One of the operational benefits of a joint approach is greater interoperability among the participating services. We addressed this in our assessment of economies of scale in O&S costs, which include the benefits of a common spare parts pool, a key part of interoperability. We did not assess other potential benefits of interoperability, such as the ability of one service to do maintenance on another's aircraft if necessary.

[3] See Mark A. Lorell, Michael Kennedy, Robert S. Leonard, Ken Munson, Shmuel Abramzon, David L. An, and Robert A. Guffey, *Do Joint Fighter Programs Save Money? Technical Appendixes on Methodology*, Santa Monica, Calif.: RAND Corporation, MG-1225/1-AF, 2013.

of joint, common programs are typically underestimated. The degree of underestimation, if any, can be used to estimate whether total costs become higher or lower.

The ultimate question we seek to answer with our full methodology is whether or not, in the end, the actual, realized cost benefits of joint aircraft programs offset and exceed any increased costs due to greater complexity, resulting in a force of aircraft with lower LCC than those of an equivalent force of specialized, single-service aircraft.

PAF sought to answer this question by assessing historical joint aircraft program outcomes and cost data from the early 1960s through today's JSF. Among the major findings was that historical joint aircraft programs have experienced so much higher rates of acquisition cost growth than single-service programs have that they have not saved overall LCC, despite any efficiencies from common efforts. Researchers also found that, nine years past Milestone B (MS B), the JSF program is not on the path to achieving the expected cost savings versus three separate notional, single-service fighter programs.

A summary of our major findings is as follows.

*Historical joint aircraft programs have experienced higher rates of acquisition cost growth than single-service aircraft programs and have not saved overall LCC.* We compared RDT&E and procurement cost-growth estimates for recent historical single-service and recent historical joint aircraft programs at the same points in their program histories following the beginning of full-scale development (MS B), adjusting for changes in procurement quantity. Cost growth was measured in dollars of constant purchasing power, so inflation effects were properly accounted for. We found that joint programs experience substantially higher cost growth in acquisition than single-service programs do.[4] Although joint aircraft programs may, in theory, save costs by sharing RDT&E resources and by achieving economies of scale in procurement and O&S, the maximum percentage theoretical savings in joint

---

[4] In general, the word *acquisition* loosely refers to the entire process of acquiring major defense systems. In our discussion of historical program costs and JSF costs, we use *acquisition* in the narrower technical sense as used by DoD cost analysts to refer to the combined RDT&E and procurement portions of a program. *Procurement* refers to the production phase of the program.

acquisition and O&S compared with equivalent single-service programs are too small to offset the historically observed additional average cost growth that joint aircraft programs experience in the acquisition phase.

*JSF is not on the path to deliver promised LCC savings.* In order to determine whether or not JSF is on track to deliver the originally promised LCC savings, we developed LCC estimates for three notional single-service fighter programs at MS B and nine years after MS B to compare with actual JSF LCC estimates at the same points in the JSF program.[5] We developed our estimates of LCC for three notional fighters based on conservative assumptions that favored JSF. We also recalculated our estimates of notional single-service fighter LCC using a different methodology to verify the robustness of our original production cost estimates. Although the JSF program was structured to overcome some of the problems encountered by past joint fighter programs, it faced the challenge of accommodating three substantially different sets of service requirements (along with international partner requirements) and ambitious technical and performance objectives (such as supersonic low observable short takeoff and vertical landing [STOVL] capability) into a single core aircraft design, with an 80-percent commonality goal among service variants. Our analysis of SAR data shows that, nine years past MS B, estimated JSF LCC are higher than if the services had pursued three separate fighter programs.

As shown in Figure S.1, at the beginning of development (using MS B program baseline data), JSF LCC theoretically would be 16 percent less than our estimate of the cost of three notional single-service fighter programs. However, the situation changes after nearly a decade. Nine years past MS B, SAR estimates of JSF LCC are higher than our LCC estimates for notional single-service fighters. As indicated by the stacked bars on the right of the figure, if we apply F-22 cost-growth percentages at MS C (9.7 years past MS B) to the notional single-service fighters' MS B estimate, then JSF LCC are 65 percent higher. JSF LCC are 37 percent higher if we calculate the notional single-service pro-

5 Nine years after MS B was the most recent point in time with full SAR cost data available for JSF at the time the analysis was undertaken.

**Figure S.1**
**Estimated Nine Years Past Milestone B, Life Cycle Cost for Joint Strike Fighter Would Be Higher Than Those for Three Notional Single-Service Programs (Assuming F-22 Cost-Growth Percentages)**

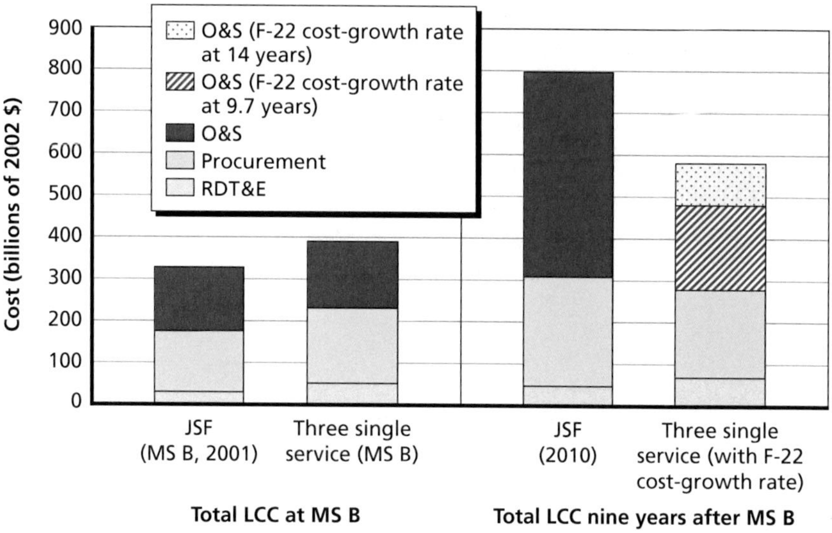


RAND *MG1225-S.1*

grams using the F-22 estimated O&S cost-growth percentage at initial operational capability (IOC), 14 years after MS B, when O&S costs were expected to be greater than in earlier estimates. Under none of the plausible conditions we analyzed did JSF have a lower LCC estimate than the notional single-service programs.

*The difficulty of reconciling diverse service requirements in a common design is a major factor in joint cost outcomes.* From the Tactical Fighter, Experimental (TFX)/F-111 program in the 1960s through the JSF program today, the attempt to accommodate multiple operating environments, service-specific missions, and differing performance and technology requirements in common joint fighter designs has increased programmatic and technical complexity and risk, thus prolonging RDT&E and driving up joint acquisition costs. At the same time, service-specific requirements and demands tend to produce less commonality and lead to more variants, thus reducing the main source of joint cost savings anticipated in procurement and O&S. Historically, joint fighter programs have typically evolved toward distinct service variants with significantly reduced commonality. For example, the congressionally mandated joint Air Combat Fighter program in the early 1970s evolved from an original goal of 100 percent commonality into two distinct platforms with zero commonality: the Air Force F-16A/B and the Navy F/A-18A/B. In other cases, necessary design compromises left the services unsatisfied and sometimes resulted in one or more partners withdrawing from the program, as in the case of the Air Force/Navy F-111 program and numerous others. These factors work against the potential for joint cost savings, which depend on maximum commonality, and are a major contributor to the joint acquisition cost-growth premium identified in our cost analysis.

*Joint aircraft programs are associated with a shrinking combat aircraft industrial base.* Looking beyond cost considerations, the pursuit of joint aircraft programs in recent decades has coincided with a reduction in the number of major fighter aircraft prime contractors from eight in 1985 to only three today. Currently, Lockheed Martin is the only prime contractor actively leading a fifth-generation manned fighter/attack aircraft development and production program (JSF) for the foreseeable future. Such a situation reduces the potential for future competition, may discourage innovation, and makes costs more difficult to control. Whether the next fighter development program is joint or single service, acquisition decisionmakers will face the challenge of a smaller industrial base and must understand the impact of acquisition strategy on the long-term health of the industry.

*Joint aircraft programs could potentially increase operational and strategic risk to warfighters.* Having a variety of fighter platform types across service inventories provides a hedge against design flaws and maintenance and safety issues that could potentially cause fleet-wide stand-downs. Having a variety of fighter platform types also increases the options available to meet unanticipated enemy capabilities. For example, during the Korean War, the U.S. Air Force was able to rapidly upgrade one of its four jet fighters, the F-86 Sabre, to meet the surprise introduction of the Russian Mikoyan-Gurevich (MiG)-15, a Soviet-designed fighter that was more capable than any other U.S. fighter in the Air Force or Navy inventory. Had the Air Force and

Navy relied exclusively on a single joint fighter other than the F-86, it might not have been able to respond quickly to the unanticipated new threat posed by the MiG-15. The more the U.S. military employs joint fighters, the fewer options will be available to meet unforeseen threats and crises in the future, as well as unanticipated safety and reliability issues that can ground entire fleets of specific aircraft types.

Informed by these findings, we recommend that, unless the participating services have identical, stable requirements, DoD avoid future joint fighter and other complex joint aircraft programs.

# Acknowledgments

We are grateful for the broad expertise, advice, constructive criticism, and assistance provided by numerous Air Force officers, as well as colleagues at RAND, without whom this study could not have been completed. Gen Donald Hoffman, as former commander of Air Force Materiel Command (AFMC), initiated this research and provided crucial guidance throughout. The authors would also like to acknowledge Lt Gen (Ret) George K. Muellner, former Principal Deputy, Office of the Assistant Secretary of the Air Force for Acquisition, and former Director and Program Executive Officer, Joint Advanced Strike Technology Program, for his invaluable insights into the early genesis of the Joint Strike Fighter (JSF) program. William Suit, Headquarters AFMC History Office, searched through vast amounts of primary source materials and provided extensive archival documentation that proved critical for our evaluation of historical joint fighter programs. Extensive information on the F-111 program derived from her careful study of the documentary record was graciously shared with us by Molly J. Waters, Operations Research Analyst, Headquarters AFMC. Former Lt Col Michael McGee, Operational Requirements Lead, JSF, Directorate of Operational Requirements, Headquarters U.S. Air Force, in the early phases of the JSF program, contributed useful information on the early parts of the process of reconciling requirements.

Jacques Gansler and Paul Kaminski, both former Under Secretaries of Defense for Acquisition and Technology, read earlier drafts of this document and provided valuable feedback. We greatly appreciate their assistance.

Many RAND colleagues contributed crucial analysis, insights, critiques, and advice in the course of this research. In this regard, we are especially grateful for the contributions of Fred Timson, James S. Chow, Obaid Younossi, John C. Graser, Donald Stevens, Daniel M. Romano, Natalie W. Crawford, Paul DeLuca, and many others. Eric Peltz, Associate Director of the RAND National Security Research Division and Director of the RAND Supply Chain Policy Center, provided a comprehensive critique of an earlier version of this document that led to significant improvements. We are particularly grateful to James S. Chow and Obaid Younossi for their formal technical reviews of the final draft of this document.

The constant support, substantive input, and critical guidance provided by Andrew R. Hoehn, RAND Senior Vice President for Research and Analysis and former Director of Project AIR FORCE; Lara Schmidt, Associate Director, RAND Project AIR FORCE; and Laura H. Baldwin, Director, Resource Management Program, RAND Project AIR FORCE, are greatly appreciated. We also owe a special debt of gratitude to our project monitor, Dr. Ross Jackson, Deputy Division Chief, Studies and Analyses Division, Headquarters AFMC, for his considerable assistance and encouragement.

## Abbreviations

| | |
|---|---|
| ACF | Air Combat Fighter (F-16) |
| AFMC | Air Force Materiel Command |
| AFTOC | Air Force Total Ownership Cost |
| A/F-X | Advanced Attack Fighter Aircraft |
| ASTOVL | advanced short takeoff and vertical landing |
| ATA | Advanced Tactical Aircraft |
| ATE | Advanced Technology Engine (F100) |
| ATF | Advanced Tactical Fighter (F-22) |
| A-X | Attack, Experimental (joint Navy–Air Force, predecessor to A/F-X, early 1990s; also the name of an Air Force program from 1966 to 1972, which then became the A-10 program) |
| CAIG | Cost Analysis Improvement Group |
| CAPE | Cost Assessment and Program Evaluation |
| CAS | close air support |
| CGF | cost-growth factor |
| CTOL | conventional takeoff and landing |
| CV | aircraft carrier capable |

| | |
|---|---|
| DoD | U.S. Department of Defense |
| EMD | engineering and manufacturing development |
| F-X | fighter, experimental; applied to many programs over the years, including the F-15 program in the late 1960s |
| FY | fiscal year |
| IOC | initial operational capability |
| JAST | Joint Advanced Strike Technology |
| JEP | Joint Engine Program (F100) |
| JIAWG | Joint Integrated Avionics Working Group |
| JPATS | Joint Primary Aircraft Training System |
| JPO | joint program office |
| JSF | Joint Strike Fighter |
| JSTARS | Joint Surveillance Target Attack Radar System |
| LCC | Life Cycle Cost |
| LIMS-EV | Logistics Installations and Mission Support–Enterprise View |
| LO | low observable |
| LTV | Ling Tempco Vought Corporation |
| MiG | Mikoyan-Gurevich |
| MS | milestone |
| NACF | Navy Air Combat Fighter (YF-17/F-18) |
| NATF | Navy Advanced Tactical Fighter |
| O&S | operations and support |
| OSD | Office of the Secretary of Defense |

| | |
|---|---|
| PAF | RAND Project AIR FORCE |
| R&D | research and development |
| RDT&E | research, development, test, and evaluation |
| SAR | Selected Acquisition Report |
| SSF | Supersonic Short Takeoff and Vertical Landing Fighter |
| STOVL | short takeoff and vertical landing |
| T1 | first production article |
| TFX | Tactical Fighter, Experimental (F-111) |
| VFX | Navy Fighter, Experimental (F-14) |

CHAPTER ONE

# Introduction

## Focus on Joint Aircraft Acquisition

The U.S. Department of Defense (DoD) has launched joint tactical fighter programs to meet the needs of multiple services many times since the 1960s. The largest and most recent program is the Lockheed Martin F-35 Joint Strike Fighter (JSF), which was developed for use by the U.S. Air Force, U.S. Navy, U.S. Marine Corps, and international partners and is currently in low-rate initial production. The primary objective of joint, versus single-service, programs is to save money by eliminating duplicate research, development, test, and evaluation (RDT&E) efforts and by realizing economies of scale in procurement and operations and support (O&S). Yet, the need to integrate multiple service requirements in a single common design increases the complexity and technical challenges of joint programs and potentially leads to higher-than-average cost growth that could reduce or even negate potential savings compared with an equivalent number of specialized single-service aircraft programs.[1]

[1] Our focus is primarily on joint fighter aircraft programs and secondarily on joint acquisition of other types of military aircraft. We do not consider joint acquisition of other types of systems, for which the outcome of the analysis could be quite different. We define *joint fighter acquisition program* as an acquisition program in which more than one service is involved from the inception in the design, development, procurement, and sustainment of the aircraft. Thus, a program in which one service unilaterally develops a fighter that is then procured jointly is viewed as a joint procurement program but not a fully joint acquisition program. For one excellent methodological discussion, including case studies on determining whether programs could benefit from a joint acquisition approach, see Thomas Held, Bruce Newsome, and Matthew W. Lewis, *Commonality in Military Equipment: A Framework*

Although there have been many analyses of the potential savings from joint aircraft programs, there have been no comprehensive analyses of actual outcomes based on cost data from historical joint aircraft programs. Several questions require detailed analysis: Have joint aircraft programs experienced greater or lesser cost growth than single-service programs? Has a joint approach yielded sufficient RDT&E, procurement, and O&S savings to offset any additional cost growth arising from the complexities of a joint approach, thus resulting in net life-cycle savings compared with single-service programs? Looking beyond cost issues, to what extent have the participating services compromised on requirements in order to achieve system commonality in joint aircraft programs? How do joint programs affect the fighter and military aircraft industrial base in the long term? Do fewer fighter aircraft designs and fewer fighter platform types across the service inventories affect operational and strategic risk for warfighters?

In theory, joint programs have the potential to save costs compared with comparable single-service programs by reducing total RDT&E expenditures and achieving economies of scale in production and O&S. But the need to accommodate different service requirements in a single design or common design family leads to greater program complexity, increased technical risk, and common functionality or increased weight in excess to that needed for some variants, potentially leading to higher overall cost and unwelcome design and performance penalties, despite these efficiencies. The fundamental question we seek to answer is whether, on average, the savings that should accrue from joint aircraft programs are sufficient to offset the additional costs arising from greater complexity and nonoptimal designs. In short, do joint fighter and other joint aircraft programs cost less overall throughout their entire life cycles than an equivalent set of specialized single-service systems?[2]

---

*to Improve Acquisition Decisions*, Santa Monica, Calif.: RAND Corporation, MG-719-A, 2008.

[2] We do not attempt to place a quantitative cost value on the effects of procuring nonoptimized joint designs but note that it could have significant effects on overall force capability.

Answering these questions will help Air Force leaders (and acquisition decisionmakers in general) to select an appropriate acquisition strategy for future combat aircraft programs and to ensure that future joint aircraft programs are carried out in the most efficient and beneficial manner possible.

## Project Objectives and Approach

In July 2011, Gen Donald Hoffman, commander of Air Force Materiel Command (AFMC),[3] asked RAND Project AIR FORCE (PAF) to conduct a detailed assessment of cost savings from joint tactical fighter programs, beginning with the Tactical Fighter, Experimental (TFX)/F-111 program in the early 1960s and ending with JSF, in order to inform acquisition strategy discussions for future combat aircraft. The analysis examined the following major questions:

- Have historical joint aircraft programs saved Life Cycle Cost (LCC) compared with comparable single-service aircraft programs?
- Is JSF on track to providing the promised LCC savings?
- What factors contributed to cost outcomes in historical joint aircraft programs, as well as JSF?
- What are the implications of a joint aircraft approach for the industrial base?
- What are the implications of a joint aircraft approach for operational and strategic risk?

We used the best data available to analyze historical aircraft programs and JSF. A major source was the RAND Selected Acquisition Report (SAR) database, which documents cost and schedule estimates and procurement quantities at different stages of more than 300 major

[3] All positions, ranks, and offices as they appear in this document are current as of the completion of the research in December 2011.

defense acquisition programs dating back to the late 1960s (the data-cutoff point for this analysis was November 2011).

However, data about historical joint fighter aircraft programs are incomplete because all joint fighter programs other than JSF that were initiated in the past 50 years did not reach the joint procurement stage.[4] Where data are lacking, we made conservative estimates (described further in Chapters Two and Three) that were generally favorable to joint programs.

Looking beyond cost considerations, we also briefly analyzed the effects of joint fighter programs on the combat aircraft industrial base, as well as operational and strategic risks associated with multiservice dependency on a single basic fighter platform design.

We adopted a two-pronged approach using quantitative and qualitative analysis to assess the potential operational and strategic risks posed by heavy dependence on a single basic platform design.[5] For the quantitative analysis, we used historical data on fleet-wide stand-downs of various Air Force fighter aircraft types, and then projected the different future probabilities of a fleet-wide stand-down of all fighters, depending on the number of different types of basic platform designs and fighter types across the inventories of the three services. For the qualitative analysis, we surveyed post–World War II U.S. air combat experience to understand how the number of different fighter types across the service inventories affected operational and strategic risk.

---

[4] Thus, JSF is not only the largest, most ambitious, and complex joint fighter program in history; it is the only fully joint fighter program (by our definition) in the past 50 years to have progressed beyond the joint development stage into the joint procurement phase.

[5] One of the operational benefits of a joint approach is greater interoperability among the participating services. We addressed this in our assessment of economies of scale in O&S cost, which include the benefits of a common spare parts pool, a key part of interoperability. The other potential benefits of interoperability, such as the ability of one service to do maintenance on another's aircraft in an emergency situation, do not affect the cost of operating the aircraft.

Greater detail on the methodologies and approaches used in the analysis reported in this document has been published in a separate document.[6]

## Organization of This Report

This report documents the major findings of the PAF analysis. Chapters Two and Three present our analysis of historical and JSF program costs, respectively. In each chapter, we describe the analytical approach; present findings for LCC, including RDT&E, procurement, and O&S; and discuss the factors that contribute to these cost outcomes. Chapter Four summarizes our conclusions regarding the implications of joint aircraft programs for the industrial base and for operational and strategic risk. Chapter Five summarizes our conclusions and recommendations for the Air Force.

---

[6] See Mark A. Lorell, Michael Kennedy, Robert S. Leonard, Ken Munson, Shmuel Abramzon, David L. An, and Robert A. Guffey, *Do Joint Fighter Programs Save Money? Technical Appendixes on Methodology*, Santa Monica, Calif.: RAND Corporation, MG-1225/1-AF, 2013.

CHAPTER TWO

# Historical Joint Fighter and Other Joint Aircraft Programs: Analysis of Savings and Costs

## Methodology

The first part of the analysis assessed cost outcomes related to actual historical joint fighter programs and other joint aircraft programs, compared with those of single-service programs. Although it is typical for major acquisition programs to experience some cost growth in the RDT&E and procurement phases, we assessed (1) whether joint aircraft programs have historically experienced higher-than-average cost growth during RDT&E and procurement and, if so, (2) whether joint programs could save enough money in RDT&E, procurement, and O&S to offset such additional cost growth, thus resulting in lower LCC compared with an equivalent number of specialized single-service aircraft.

We began by identifying and examining 11 major joint fighter programs (Table 2.1) that were proposed from 1962 through 1996 (i.e., prior to JSF). All were intended to be major acquisition programs, and most were initiated by DoD or Congress.[1] Of these, seven programs never progressed beyond the proposal stage to actual development and

---

1 F-117N/AF-117X and the Joint Advanced Strike Technology (JAST) program are exceptions. F-117N and AF-117X were largely company initiatives promoted by Lockheed. When JAST was first launched, it was explicitly described by DoD as a joint technology demonstration and maturation program, not a joint acquisition program. However, it rapidly evolved into a de facto acquisition program and was explicitly recognized as such when it was reorganized and renamed as the JSF program in 1996.

It is important to note that Table 2.1 shows original program intent, not actual outcomes of these programs.

**Table 2.1**
**Actual and Proposed Pre–Joint Strike Fighter Joint Tactical Fighter Programs Reviewed**

| | | | Design and Technology Development | | EMD | | Procurement | |
|---|---|---|---|---|---|---|---|---|
| Program | Lead Service | Dates | Single Service | Joint | Single Service | Joint | Single Service | Joint |
| TFX/F-111[a] | Air Force | 1962–1968 | | X | | X | | X |
| F-4[a, b] | Navy | 1962–1979 | X | | X | | | X |
| A-7[a, b] | Navy | 1965–1984 | X | | X | | | X |
| F-X, VFX, ATE, JEP[c] | Air Force | 1966–1971 | | X | | X | | X |
| F-15N[c, d] | Air Force | 1971, 1973 | X | | | Partial | | X |
| ACF, NACF (YF-16, YF-17)[c, d] | Air Force | 1974–1975 | X | | | X | | X |
| ATF, NATF, JIAWG[c, d] | Air Force | 1986–1991 | X | | | X | | X |
| ATA[b, c] | Navy | 1986–1991 | X | | | X | | X |
| A-X, A/F-X[c] | Navy | 1991–1993 | | X | | X | | X |
| F-117N, AF-117X[c, d] | Air Force | 1993–1994 | X | | | X | | X |
| JAST[a] | JPO | 1993–1996 | | X | | X | | X |

[a] Progressed beyond design proposal and technology risk-reduction stages.

[b] U.S. Navy baseline design and RDT&E.

[c] Proposal only.

[d] U.S. Air Force baseline design and RDT&E.

NOTE: The table reflects original program intent. EMD = engineering and manufacturing development. F-X = fighter—experimental; evolved into the F-15 program. VFX = Navy Fighter, Experimental (F-14). ATE = Advanced Technology Engine (F100). JEP = Joint Engine Program (F100). ACF = Air Combat Fighter (F-16). NACF = Navy Air Combat Fighter (YF-17/F-18). ATF = Advanced Tactical Fighter (F-22). NATF = Navy Advanced Tactical Fighter. JIAWG = Joint Integrated Avionics Working Group. ATA = Advanced Tactical Aircraft. A-X = Attack, Experimental (joint Navy–Air Force, predecessor to A/F-X, early 1990s). A/F-X = Advanced Attack Fighter Aircraft. JPO = joint program office.

procurement, though they did include many detailed proposals and estimates of costs, some of which we were able to examine. Two programs, the F-4 and the A-7, resulted in joint production and deployment, but we do not consider them to be true joint programs because they were developed as single-service programs by the Navy. The TFX/F-111 emerged from a joint Air Force/Navy requirement and design effort and progressed four years into RDT&E as a joint development program, at which time the Navy withdrew because of dissatisfaction with weight and cost growth and the performance of the Navy F-111B prototype. Subsequently, the Navy launched its own single-service program, which led to the F-14 fighter. Although begun as a technology demonstration and maturation program, JAST is included as a joint aircraft program because it evolved into a de facto joint acquisition program and was the immediate predecessor to the JSF program.

Although there are significant qualitative and proposal data available on joint fighter programs prior to JSF that are worthy of study and analysis and that provide extensive insight into the challenges and issues posed by joint fighter programs, there is a lack of cost data derived from actual program outcomes. Full and comparable cost data from 1960s joint fighter programs, such as TFX/F-111, F-4, and A-7, are incomplete (because none of these programs ended up being fully joint), not fully available in SARs, and of uncertain quality.[2]

In order to make cost-growth comparisons using actual program cost estimates and cost data with extensive budgetary time series, we broadened the analysis to include military aircraft other than fighters, for which there are substantial SAR data. We included all recent unclassified aircraft major defense acquisition programs from the mid-1980s on with a significant time series of data (nine years or more beyond Milestone B [MS B]) and that included a major development effort. We analyzed four joint programs (JSF, the Joint Primary Aircraft Train-

---

[2] SAR reporting did not yet exist in the early 1960s and became more reliable and comprehensive only in the 1980s. As noted earlier, prior to JSF, not one of the fully joint fighter proposals (by our definition) progressed into the production phase, and only the TFX/F-111 experienced a period of actual joint development. In the case of the TFX/F-111, some data are available from the joint RDT&E phase, during which the Navy was still participating, but not in the current SAR format.

ing System [JPATS] T-6A, the Joint Surveillance Target Attack Radar System [JSTARS] E-8, and the V-22 Osprey) and four single-service programs (C-17, F/A-18E/F, F-22, and T-45 Training System). All cost comparisons were adjusted for changes in quantity, and the analysis was done in dollars of constant purchasing power to remove the effects of inflation.[3]

## Major Cost Findings

First, we analyzed the RDT&E and procurement (together called the *acquisition phase*)[4] cost estimates for each program from five years after MS B to a point nine years past MS B.[5] Figure 2.1 shows that, at every year during this period, joint aircraft programs on average experi-

---

3 We recognize that the data set for this part of the analysis is relatively small. Our conclusions are based on the best data available for all major defense acquisition aircraft programs (excluding "black" or highly classified programs, such as the B-2), both joint and single service, from the mid-1980s to the present, that have significant acquisition actual data time series in the SARs, and that included substantial RDT&E. Although the data vary from program to program, the pattern is clear: The single-service programs, with a cost-growth range at nine years beyond MS B of –2 percent to 48 percent, are either lower or close to the low end of the joint program range of 40 percent to 97 percent cost growth. In addition, there are no outlying cases that drive the average cost growth shown here.

4 In general, the word *acquisition* loosely refers to the entire process of acquiring major defense systems. In our discussion of joint historical program costs and JSF costs, we use *acquisition* in the narrower technical sense, as used by DoD cost analysts to refer to the combined RDT&E and procurement portions of a program. *Procurement* refers to the production phase of the program.

U.S. Department of Defense, *Operation of the Defense Acquisition System*, Instruction 5000.02, December 8, 2008, structures major acquisition programs into five main phases. The second milestone event, called the MS B review, assesses the program's readiness for entry into the EMD phase. During this phase, the item is fully developed into a system that, after successful completion of the MS C review, enters the production phase.

5 This data range from these program years was chosen for the following reasons. Five years after MS B was chosen as the beginning point because, at fewer than five years past MS B, substantial cost growth is not yet recognized in some programs. Nine years after MS B was chosen as the end point because this was the furthest point beyond MS B that we had reasonably complete sets of cost data for all eight aircraft programs at the time of the data-cutoff point for this research, which was November 2011.

**Figure 2.1**
**Historical Joint Aircraft Programs Have Incurred an Acquisition Cost-Growth Premium Compared with Single-Service Aircraft Programs**

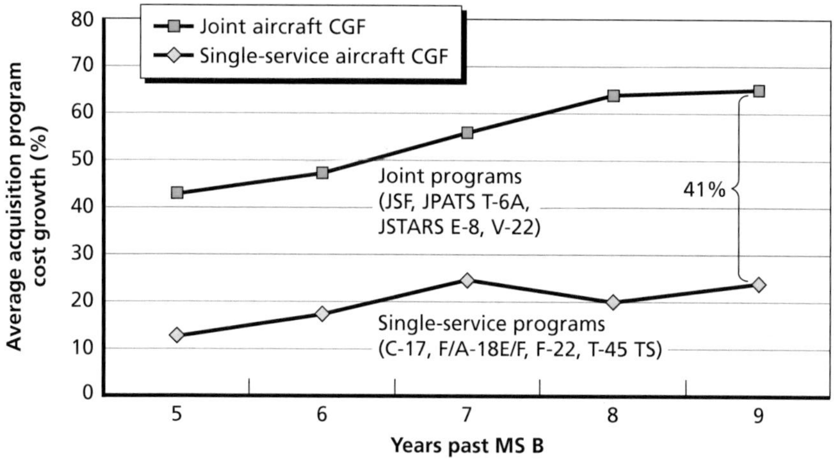


SOURCE: Authors' analysis of SAR cost data.
NOTE: CGF = cost-growth factor.
RAND *MG1225-2.1*

enced considerably more average acquisition program cost growth than single-service aircraft programs (to which we will refer as a joint cost-growth *premium*) did. At the point furthest beyond MS B for which we had full data (nine years past MS B), acquisition cost estimates had grown by an average of 24 percent for the single-service programs and 65 percent for joint programs.

This comparison indicates that, on average, historical joint aircraft programs incurred an additional 41 percent of cost growth in the acquisition phase at a point nine years after MS B. However, Figure 2.1 also shows that the percentage size of the average joint aircraft cost-growth premium varied over this period from a low of about 30 percent to a high of about 44 percent. At no point did the average cost growth for joint aircraft acquisition programs exceed that for single-service programs by less than 30 percent. (The reasons for cost growth are discussed in the next chapter.)[6]

Next, we assessed whether the theoretical maximum savings that could accrue from joint RDT&E and procurement were sufficient to offset the joint acquisition cost-growth premium described above, thus yielding a net acquisition savings over single-service programs. We used algebraic equations to determine the maximum potential savings that a notional "ideal" joint program could, in theory, achieve in RDT&E and procurement. For example, in a joint program in which two services acquire equal numbers of 100-percent common fighters and pay an equal share of all costs, each service will save 50 percent in RDT&E and 13 percent in procurement compared with two single-service programs (assuming a typical fighter production cost improvement curve with an 87-percent slope). In a representative modern fighter aircraft program, in which production or procurement spending is four times that of RDT&E, the overall joint acquisition savings would be on the order of 20 percent, as shown by the first two bars on the left in Figure 2.2. For the purpose of comparison, we consider this to be representative of the theoretical maximum cost savings for a representative joint fighter aircraft acquisition program.[7]

However, even if a notional historical two-service joint aircraft program realized the theoretical maximum 20-percent savings in acquisition costs, it would still end up costing more than two comparable separate single-service programs after cost growth because of the 41-percent average joint cost-growth premium discussed above. The two bars on the right in Figure 2.2 compare the total acquisition costs of a notional two-partner joint program with two comparable separate

6 For a more detailed discussion of this comparison and its methodology, see Lorell, Kennedy, et al., 2013, Appendixes B and C.

7 The 20-percent figure is based on the relation $(0.50 \times 0.20) + (0.13 \times 0.80) = 0.204$, or 20.4 percent. Here, 0.50 is the research and development (R&D) saving, 0.13 is the production saving, and 0.20 and 0.80 are the R&D and procurement shares of a typical single-service program, respectively. The theoretical maximum savings will vary depending on the relative percentage size of each of the two components (RDT&E and procurement) compared with the overall total acquisition costs and the relative number of aircraft procured of each variant.

**Figure 2.2**
**Theoretical Maximum Savings from a Representative Two-Service Joint Aircraft Acquisition Program Are Offset by the Average Joint Acquisition Cost-Growth Premium**

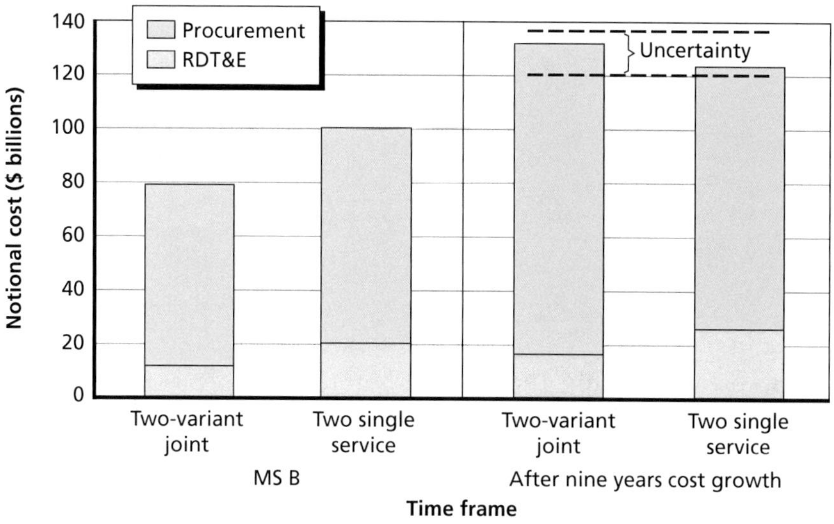


NOTE: Cost growth applied to the single-service program is the average at nine years past MS B. Uncertainty reflects the range of joint program cost-growth premiums in years 5 through 9 past MS B.
RAND *MG1225-2.2*

single-service programs after nine years of cost growth. After applying the average cost growth for historical joint and single-service aircraft programs to the baseline RDT&E and procurement estimates, we find that the joint acquisition program now costs 6 percent more than the single-service programs.[8]

However, as noted in our discussion of the joint cost-growth premium shown in Figure 2.1, the size of the joint cost-growth premium

[8] The 41-percent average joint cost-growth premium applies to the overall acquisition program cost comparisons and is a weighted average of a separate average joint RDT&E cost-growth premium of 43 percent (68 percent average joint RDT&E cost growth versus 25 percent average single-service RDT&E cost growth) and a 41-percent joint procurement cost-growth premium (64 percent average joint procurement cost growth versus 23 percent average single-service procurement cost growth), which were applied separately in the calculations for this figure.

varies over time. To reflect the range of joint cost-growth premiums in the years from five through nine years past MS B, we include an uncertainty band on the notional two-service joint fighter acquisition cost estimate bar shown in Figure 2.2. This uncertainty band overlaps both acquisition cost estimates for a two-service joint fighter program and two single-service fighter programs at a point nine years past MS B. This suggests that, after nine years of cost growth, the total acquisition cost of an "ideal" two-service joint program could be expected to be essentially the same as the total acquisition cost of two separate single-service fighter acquisition programs. In other words, a two-service joint fighter program would not save acquisition costs versus two comparable single-service fighter programs.

We next assessed whether joint aircraft programs could save enough O&S costs to offset the effects of the joint aircraft acquisition cost-growth premium.[9] The O&S savings required depends on the proportion of LCC estimated to be spent on RDT&E, procurement, and O&S in the program baseline estimate at MS B. If O&S makes up a very large share of expected total program LCC, then less O&S cost savings are required to offset the average joint aircraft acquisition cost-growth premium. The smaller the O&S share, the greater the O&S savings required. For the purpose of this analysis, we assume a representative fighter aircraft program that allocates 54 percent of LCC to acquisition and 46 percent to O&S.[10] Using this assumption, we calculate that 10.3 percent savings in joint O&S costs would be needed to offset the joint acquisition cost-growth premium, assuming that the theoretical maximum representative joint fighter savings in RDT&E and procurement are realized, the average joint acquisition program cost-growth premium at nine years after MS B is applied, and there is no O&S cost growth.

---

[9] Unfortunately, our SAR database did not include sufficient data on O&S cost estimates to compare O&S cost growth on the joint versus single-service aircraft programs we examined. Therefore, we assumed zero cost growth nine years after MS B for both joint and single-service aircraft programs. For our cost comparison of JSF with three notional single-service fighters discussed later in this document, we did have sufficient O&S cost data to assess O&S cost growth.

[10] Derived from the JSF MS B baseline estimate.

How much O&S savings are joint fighter programs likely to actually achieve under ideal circumstances? This question is difficult to answer directly because none of the fully joint historical fighter programs identified in Table 2.1 has been jointly fielded and operated.[11] Moreover, SAR data are not available to conduct the same kind of cost-growth comparisons for joint versus single-service aircraft program O&S cost growth that we carried out for the RDT&E and procurement phases.[12]

Instead, we developed a methodology using actual O&S data from current Air Force fighters to assess the O&S savings that would accrue to an "ideal" joint fighter program, in which the participating services procured identical types of common fighters and used a common support infrastructure and approach in those O&S areas affected by economies of scale. To do this, we analyzed Air Force Total Ownership Costs (AFTOCs) and Logistics Installations and Mission Support–Enterprise View (LIMS-EV) databases using regression analysis to determine how actual O&S costs of Air Force fighters vary with different fleet sizes. This approach allowed us to simulate ideal cases in which two or three services operate equal numbers of identical fighters and to observe how economies of scale would affect O&S costs.[13] This approach favors joint programs because the aircraft type mix, force

---

[11] This statement, of course, excludes the F-4 and A-7. As noted earlier, these are not true joint programs by our definition because they were developed unilaterally by a single service and then only procured jointly. In addition, appropriate O&S data on the Navy and Air Force versions of the F-4 and A-7 are not available. Furthermore, it appears that spares pooling, common depots, and other forms of joint O&S were not attempted on these programs.

[12] SARs have not included consistent O&S cost estimates and categories from early phases of programs until relatively recently.

[13] We examined data for A-10s, F-15s, F-16s, and F-22s from 1996 through 2010. Each Air Force fighter type analyzed includes numerous different variants and blocks. However, our analysis kept the same relative mix of numbers of existing variants of each fighter type when we tested for the effect of changing the size of the force structure on O&S costs for that type of fighter. Thus, we compared the effect on O&S costs of changing the number of common fighter types while keeping the existing percentage mix of the variants constant. Because all the aircraft are Air Force fighters, which are operated and supported by an identical Air Force O&S infrastructure, this approach simulates the O&S cost savings that could be achieved in an ideal joint program made up of common aircraft types across the services using an identical O&S infrastructure.

structure, and O&S infrastructure are assumed to be 100 percent identical across the services (this would not likely be the case in a real joint fighter program).

Figure 2.3 shows the maximum O&S savings for an ideal notional two-service joint fighter program, based on this analysis of O&S data. We distinguish between overall O&S cost savings and savings in categories in which one could expect to see economies of scale (e.g., depot-level reparables and aircraft overhaul, but not fuel).[14] We define an ideal notional two-service joint fighter program as one in which each service procures an equal number of the same types of aircraft, and all O&S activities that are affected by economies of scale are con-

**Figure 2.3**
**Maximum Operations and Support Savings from Ideal Joint Fighter Programs Are Unlikely to Offset the Joint Aircraft Acquisition Cost-Growth Premium**

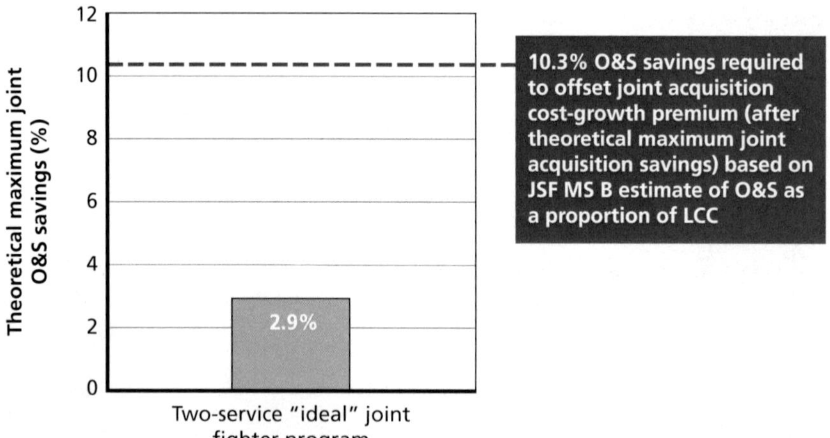


NOTE: *Ideal* here refers to having an equal number and mix of aircraft per service and 100 percent commonality in O&S activities and infrastructure.
RAND *MG1225-2.3*

[14] We segregated out O&S categories using standard AFTOC and Office of the Secretary of Defense (OSD) Cost Analysis Improvement Group (CAIG; now Cost Assessment and Program Evaluation, or CAPE) O&S cost categories. Our assessment indicated that about two-thirds of fighter O&S costs could theoretically be affected by economies of scale.

ducted 100-percent jointly and in common. Joint O&S cost savings for an ideal two-service joint fighter program would be 4.6 percent in categories affected by economies of scale and 2.9 percent overall. These savings fall short of the 10.3-percent O&S savings required to offset the historical average joint aircraft acquisition cost-growth premium at nine years past MS B, as discussed above. Of course, more than 10.3-percent savings in joint O&S costs would be necessary for a joint program under these assumptions to cost less than two comparable single-service programs.[15]

Using this analysis, we conclude that, although joint aircraft programs do, in theory, save costs by sharing RDT&E resources, increasing production runs, and utilizing economies of scale in O&S, these savings are too small to offset the substantial additional average cost growth historically observed in the acquisition phase. Historical joint aircraft programs have not yielded overall LCC savings compared with single-service programs.

## Factors Contributing to Cost Growth

An important factor contributing to this joint aircraft acquisition program cost-growth premium is the tension between the need to attain maximum design and system commonality, which is the basis of potential joint cost savings, and service-specific requirements, which tend to reduce commonality. Historically, the services have entered joint aircraft programs with unique requirements that arise from differences in operating environments, missions, doctrine, and operational concepts. Launch and recovery of fighters on an aircraft carrier involve funda-

---

[15] As noted in the discussion of Figures 2.1 and 2.2, the average joint aircraft acquisition cost-growth premium varies between five and nine years after MS B. Using an uncertainty band reflecting this variation in the joint acquisition cost-growth premium, the percentage savings in joint O&S costs necessary to offset the joint cost-growth premium to reach LCC parity with two single-service fighter programs varies between 15.4 percent and –5.1 percent. This uncertainty band suggests that, nine years after MS B, LCC of a two-service joint program can be expected to be essentially equivalent with the total of two separate single-service programs. For a more complete explanation of the methodology and data used in this analysis, see Lorell, Kennedy, et al., 2013, Appendixes B and C.

mentally different design challenges from those of takeoff and landing on a land base.[16] Short takeoff and vertical landing (STOVL) capability imposes even greater demands than conventional land- and carrier-based aircraft, particularly regarding air vehicle weight, engine thrust requirements, and the complexity of the propulsion system. Attempts to reconcile different requirements in a common-core airframe increase technical complexity and risk, which can lead to RDT&E and procurement cost growth. At the same time, efforts to optimize variants for each service decrease commonality, thus reducing the potential for procurement and O&S cost savings. The increased complexity and decreased commonality can lead to greater cost growth than for single-service programs, especially when initial baseline cost estimates have assumed optimistic projections of theoretical maximum savings from commonality, as has often been the case historically.[17]

Figure 2.4 illustrates the tensions between commonality and service optimization in four historical joint fighter programs from the 1960s and 1970s, each of which began with the goal of 100-percent commonality but diverged into unique service variants. For example, the TFX program began as a joint program for a multirole fighter with

---

[16] Our assessment of the data available on the 11 proposed and actual historical joint fighter programs listed in Table 2.1 indicates that nearly all historical Air Force design proposals that were later modified for maximum commonality with a common carrier–based Navy fighter included a roughly 2,000- to 3,000-pound empty-weight penalty to accommodate the more robust airframe structure, landing gear, and other design factors necessary for catapult launch and carrier deck recovery on an aircraft carrier. Other requirements for carrier fighters, such as low landing approach speed, high controllability at slow approach speeds, excellent forward visibility at high angles of attack, "wind over deck" factors, carrier deck "spotting" factors, and deck elevator load limitations, can have significant effects on air vehicle design, weight, maneuverability, and other performance factors.

[17] We examined the original internal cost savings calculations made by the OSD Systems Analysis office for the TFX program, and they appear to assume an "ideal" program with 100-percent commonality between Air Force and Navy variants and no additional costs for the greater developmental complexities involved. (However, it was never assumed that the Air Force and Navy would buy equal numbers of TFX aircraft.) Although the data are less complete and not in the same format as modern SARs, the development phase of the F-111 program later appears to have experienced more than 100-percent cost growth during RDT&E compared with an average of about 30 to 40 percent cost growth for major military defense programs during the 1950s and 1960s.

**Figure 2.4**
**Differing Operational Environments and Service Requirements Led to Many Variants, Less Commonality, and Greater Program Complexity, and Thus Reduced Joint Cost Savings**

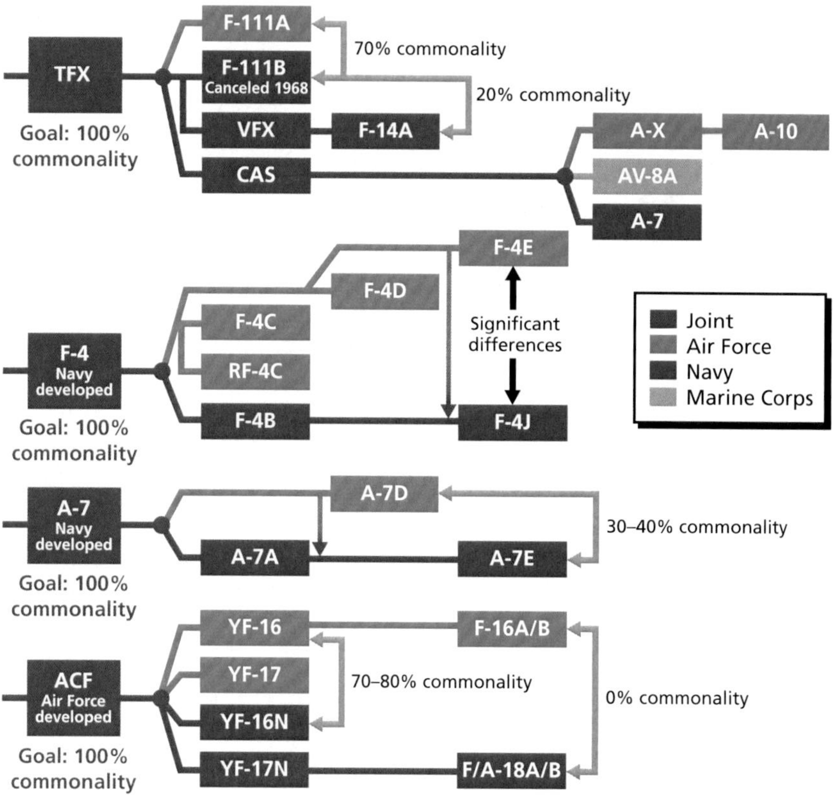


RAND *MG1225-2.4*

100-percent commonality for all participating services; however, it quickly evolved into the F-111A variant for the Air Force and the F-111B variant for the Navy. In an attempt to keep the aircraft as common as possible, each service had to accept design penalties to accommodate the other service's requirements. These penalties became so significant from the Navy perspective that the Navy eventually withdrew from the program before the completion of development and launched its own single-service program for the F-14. The original TFX close air support (CAS) mission requirement dropped out of the design very early in the program because of insurmountable design and cost issues, and it eventually evolved into the A-7A developed unilaterally by the Navy and later procured jointly by the Air Force as the A-7D variant. The Air Force A-7D and later Navy A-7E variant evolved to meet unique service requirements and had only about 30 to 40 percent production-floor commonality. The F-4 and ACF programs had similar experiences, with the latter resulting in two aircraft with no commonality: the F-16A/B for the Air Force and F/A-18A/B for the Navy.[18]

These examples illustrate the persistent tension between the need to maximize system commonality to achieve the greatest cost savings possible and the difficulty of reconciling differing service requirements, which has historically worked against the realization of theoretical joint cost savings. In the next chapter, we examine how JSF, as the latest joint fighter program, has dealt with this tension and analyze the program's potential for joint cost savings.

---

[18] The F-4A/B and the A-7A were originally developed entirely by and for the Navy. After development, the Secretary of Defense mandated Air Force procurement of both of these fighter/attack aircraft but permitted modification of the aircraft to meet unique Air Force requirements. The ACF program included a competition between the YF-16 and YF-17 prototypes. Congress mandated that the Navy would buy a navalized version of whichever fighter won the competition (the YF-16N or YF-17N). The YF-16 won the competition and was developed and procured by the Air Force as the F-16A/B. The Navy, however, argued that neither prototype met its requirements and instead developed the YF-17N into the F/A-18A/B, a fighter that had zero commonality with the Air Force F-16.

CHAPTER THREE

# Joint Strike Fighter Program: Analysis of Savings and Costs

The officials who established the JSF program were aware of the problems encountered in previous joint fighter programs, and they sought innovative approaches to address them. For example, JSF officials structured a JPO that was intended to counter service parochialism and establish a common set of requirements through extensive cost and performance trade-off analyses that would meet all participating services' threshold requirements.[1] Engineers and acquisition experts focused on affordability as a top-line key performance parameter. Between 1995 and 2000, there was a series of detailed cost/benefit design trade-off analyses and an iterative annual Joint Interim Requirements Document intended to move incrementally toward maximum system commonality while meeting the most-important individual service requirements in order to ensure full service buy-in into a final Joint Operational Requirements Document. Development of specific service variants was intended to encourage the services to support the compromises necessary to achieve a common baseline design with an overall objective of 80-percent commonality among the three variants. At the same time, JSF was the first joint fighter program in history to include all three services that employ fixed-wing fighter/attack aircraft. Yet, the inherently difficult goal of achieving maximum commonality while meeting three services' dif-

[1] This was a radically new approach. At least ten of the 11 historical joint fighter programs we examined began with an original design concept optimized for a single service and envisioned modification of this baseline design to meet the requirements of the second participating service.

ferent requirements (as well as foreign participants) proved to be an enormous challenge, with serious implications for program costs.

## Methodology

Our major goal in analyzing the JSF program was to determine in a quantitative fashion whether the program should be expected to achieve overall LCC savings compared with an equivalent set of notional single-service fighter programs. Table 3.1 summarizes the methodology used to estimate these costs. First, we used SARs to determine JSF costs and cost growth between MS B (documented in the December 2001 SAR) and approximately nine years past MS B (documented in the December 2010 SAR, the latest available SAR at the time this analysis was completed in November 2011). Costs were in constant dollars, and aircraft quantities were adjusted to make fair comparisons between the two SAR estimates.

Analyzing single-service program costs for comparison is more challenging and complex than assessing those for JSF because it requires one to infer the path not taken. We developed representative cost estimates of what three single-service programs might look like. Here, as in our quantitative analysis of historical joint aircraft programs, we made conservative assumptions that generally favored the joint program approach.

To estimate baseline RDT&E, we accepted the JSF JPO and contractor's estimate that JSF would be able to carry out the RDT&E of three different fighter variants for 60 percent of the cost of three single-service programs combined.[2] Thus, RDT&E for three single-service

---

[2] JPO and contractor estimates varied over time. In 1998, three years prior to MS B, JPO documentation said that RDT&E costs for single-service programs would be twice JSF RDT&E costs. By the early 2000s, this had been adjusted to the estimate that RDT&E for three single-service programs would be 1.5 times the cost of JSF RDT&E. One publicly available JPO briefing from 2008 shows that RDT&E for three single-service programs would be 1.67 times JSF RDT&E cost, which is the figure we use in this analysis. See, for example, Maj Gen Charles R. Davis, program executive officer, F-35 Program Office, "JSF Production," briefing presented at the Aviation Week Aerospace and Defense Finance Conference, New York, November 2008.

**Table 3.1**
**Methodological Approach for Comparing Actual Joint Strike Fighter Costs with Those of Three Notional Single-Service Fighters**

| Cost | JSF | Three Single-Service Programs |
|---|---|---|
| Baseline | Official MS B estimate from December 2001 SAR | RDT&E assumes that three fighters cost 67% more than JSF (per JPO estimate).<br><br>Procurement assumes three programs, with JSF MS B quantities by variant:<br>• No shared learning, but same T1 costs, aircraft weights, and learning curves (by variant) derived from JSF baseline estimate and funding profiles<br>• No weight/cost/performance benefit from single-service design optimization<br>• O&S assumes penalty over JSF baseline for decreased economies of scale (zero commonality), derived from AFTOC regression analysis. |
| Cost growth since MS B | December 2010 SAR estimates adjusted to MS B quantity and foreign partners | RDT&E and procurement cost-growth percentages are same as F-22 program at same point (~9 years past MS B).<br><br>O&S F-22 cost-growth percentages at both MS C (9.7 years past MS B) and IOC (14 years past MS B) |

NOTE: T1 = first production article. IOC = initial operational capability.

programs would be 67 percent higher than the RDT&E estimate for JSF at MS B.[3]

To estimate baseline procurement, we assumed three separate programs with no commonality. Each service procured the same number of aircraft as the corresponding JSF service variant planned at MS B. We penalized the notional single-service fighters by assuming no shared learning or commonality.[4] The notional single-service fighters had no T1 weight or cost advantages over the actual individual service variants

[3] Expressed algebraically, RDT&E for three single-service fighters ($x$) equals 1.67 times that for JSF ($y$), or $x = 1.67y$. Thus, the percentage of $x$ represented by $y$ is $\left(y/x\right) = \left(1/1.67\right)$, or 60 percent.

[4] The original single-service fighter concepts assumed some commonality. For example, the original concept for the single-service Air Force MRF was to share the same engine with the single-service Navy fighter, the A/F-X. This engine was originally planned to be the Pratt and Whitney F119, the same engine used on the F-22. Thus, engine commonality and economies

of JSF. We did this even though a single-service fighter developed by and optimized exclusively for a single service's requirements might have been able to eliminate design, weight, cost, and performance penalties incurred due to meeting other services' requirements.[5] Individual production improvement curves were used for each of the single-service programs, derived from the relevant variants in the JSF program.[6]

To estimate baseline O&S costs for notional single-service fighters, we assumed that zero commonality would eliminate the joint economies of scale (though, in reality, single-service programs could benefit from some common elements, such as engines, as had been originally planned in the predecessor single-service programs). Thus, a penalty was applied based on our analysis of maximum O&S savings for an ideal joint program derived from the AFTOC and LIMS-EV databases.

To estimate the notional single-service program costs at nine years past MS B (to correspond to the JSF 2010 SAR), we again used conservative assumptions that would tend to favor JSF. Actual F-22 cost-growth rates were used for the RDT&E and procurement phases.[7] O&S costs usually make up the single largest part of overall LCC but are difficult to estimate accurately before full development and

---

of scale could have been achieved with the F-22, as well as with the Navy A/F-X and Air Force MRF.

5 The argument here is that all three JSF design variants are likely to have design penalties from each user service's perspective caused by the need to maximize system commonality while meeting all three services' mission requirements and while operating in three distinct operational environments. Thus, a single-service Air Force conventional takeoff and landing (CTOL) aircraft designed exclusively by and for the Air Force might have either weighed less or had greater capabilities for meeting Air Force–specific mission requirements at the same or greater empty weight.

6 As inferred from the SAR MS B JSF cost estimate and annual funding profiles, as well as commonality levels.

7 We view this as a very conservative approach. The F-22 is a heavier, much larger, much higher-performance fighter than the JSF Air Force variant. It was a relatively high-risk development program because it was the first supersonic/super cruise low observable (LO) fighter ever developed. It was also the first fighter with a fully integrated avionics system. From the very beginning of RDT&E, the F-22 experienced significant cost growth and was a target of extensive criticism from Congress, the Government Accountability Office, and the press.

deployment. For robustness in the estimates, we applied two O&S cost-growth rates. One was the rate of F-22 O&S cost growth at MS C (9.7 years past MS B, roughly where JSF is today). The other was the rate of F-22 O&S cost growth at IOC (14 years past MS B).[8]

## Major Cost Findings

The JSF program was initially envisioned as an affordable program costing $175 billion (fiscal year [FY] 2002 dollars) for procurement of 2,852 joint fighters, as shown on the left of Figure 3.1.[9] The column on the right represents RDT&E and procurement costs for the same number of notional single-service fighters at MS B. Here, one sees the expected theoretical advantages of a joint program: The single-service programs have 67 percent higher RDT&E, and they do not benefit from longer common learning or improvement curves. Thus, comparing the actual baseline acquisition estimate for JSF as of MS B (October 2001) with our baseline estimate for three equivalent single-service fighters, the JSF estimate shows an overall acquisition cost savings of 25 percent.

Figure 3.2 shows the RDT&E and procurement costs for JSF and notional single-service programs at nine years past MS B. The JSF column represents the actual cost estimate data from the December

---

[8] Historical SAR data show that, in general, the further beyond MS B a fighter program progresses, the more realistic (and higher) the O&S cost estimates become. A more detailed discussion of this methodology is available in Lorell, Kennedy, et al., 2013, Appendix E.

[9] All cost figures in this section are 2002 dollars, consistent with the baseline JSF estimate from the December 2001 SAR. Although the number of aircraft declined considerably shortly after MS B, with the Navy cutting back its procurement plans, we normalize the cost estimates in this section to 2,852 aircraft for consistent comparisons. The JSF baseline estimate also includes the improvement curve benefit of an expected UK procurement of 150 F-35B STOVL variants, per the December 2001 SAR. The baseline MS B estimate assumes a $2 billion UK contribution to RDT&E, but the RDT&E estimate shown includes only estimated U.S. expenditures. The STOVL single-service fighter is also assumed to benefit from a $2 billion UK contribution to RDT&E because the predecessor advanced short takeoff and vertical landing (ASTOVL) program had been a combined U.S.-UK program since 1983.

**Figure 3.1**
**Estimated at Milestone B, Research, Development, Test, Evaluation, and Procurement for Joint Strike Fighter Would Cost Less Than for Three Notional Single-Service Programs**

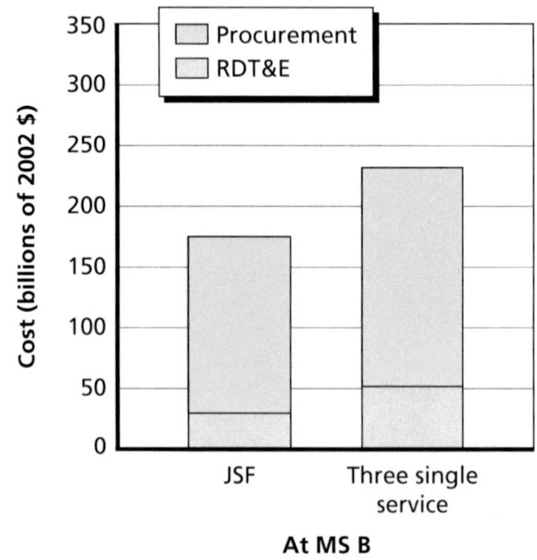


RAND *MG1225-3.1*

2010 SAR, adjusted for foreign contributions and normalized to the original planned MS B procurement numbers. This estimate also assumes foreign partner procurement of 730 JSF variants, including the associated cost improvement curve benefits, but it includes only direct U.S. costs.[10] For the notional single-service programs, we apply the rate of cost growth experienced by the F-22 over approximately the same time period.[11] JSF acquisition costs would be 10 percent higher than those for the notional single-service programs. Similar to our

[10] The 2010 SAR assumed greater foreign participation (730 STOVL and CTOL variants combined) than the 2001 SAR (150 STOVL variants).

[11] The single-service programs, of course, do not benefit from any commonality and the associated longer production runs. Nor are they given credit for any foreign sales other than the 150 UK STOVL variants. This is justified based on the fact that, prior to JSF, the single-service ASTOVL program dating from 1984 had always been an international UK-U.S. effort and assumed U.S. and UK procurement.

**Figure 3.2**
**Estimated Nine Years Past Milestone B, the Joint Strike Fighter Acquisition Program Would Cost More Than Three Notional Single-Service Acquisition Programs (with F-22 Cost-Growth Percentages)**

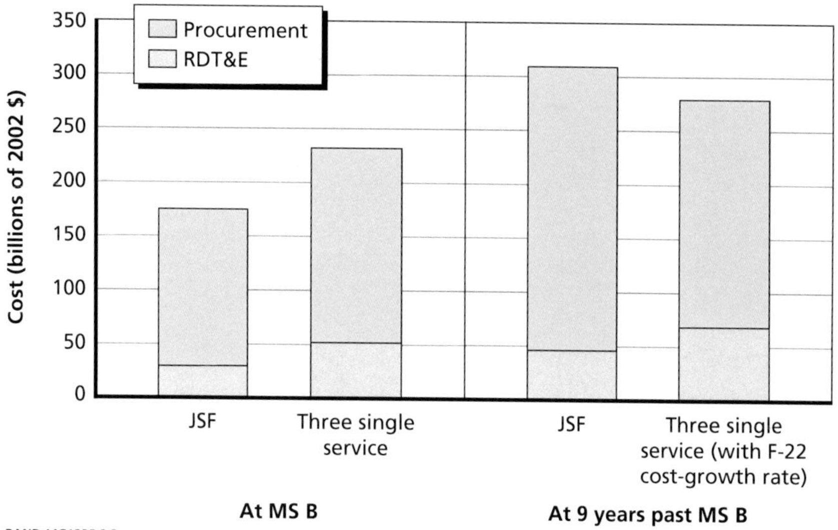


RAND *MG1225-3.2*

finding for historical joint fighters, JSF experiences higher acquisition cost growth than notional single-service programs, sufficient to cancel out the cost advantage it was expected to enjoy at MS B.

Our analysis of total LCC (RDT&E, procurement, and O&S) shows a similar pattern to that for acquisition costs (RDT&E and procurement alone). As shown in Figure 3.3, at MS B, JSF is expected to save 16 percent in total LCC over notional single-service programs when the expected economies of scale in O&S are combined with the expected RDT&E and procurement cost benefits discussed above.

However, as shown in Figure 3.4, the O&S estimate for JSF (normalized for original procurement quantities) grew substantially in the nine years between MS B and 2010. Compared with the notional single-service programs (assuming F-22 O&S cost-growth percentages at 9.7 years past MS B), the JSF total LCC is 65 percent higher. Even if we assume that notional single-service fighters experience the cost-growth rates estimated when the F-22 reached IOC (14 years after

**Figure 3.3**
**Estimated at Milestone B, Life Cycle Cost for the Joint Strike Fighter Would Be Less Than Those of Three Notional Single-Service Programs**

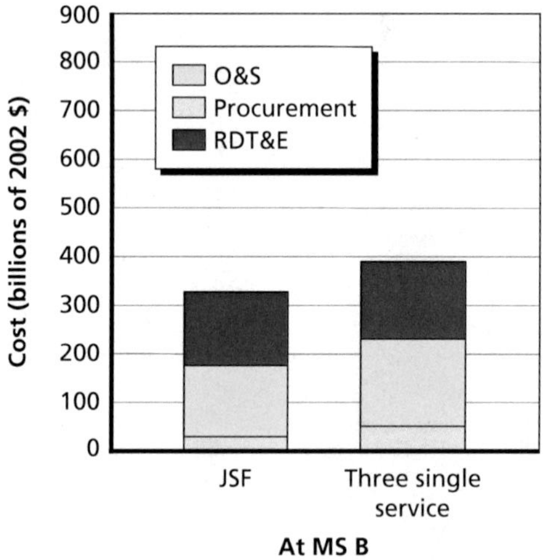


RAND *MG1225-3.3*

MS B), the JSF LCC estimate is still 37 percent higher.[12] Under none of the plausible conditions we analyzed did JSF have a lower LCC than the notional single-service programs at nine years past MS B.[13]

---

[12] We also note that a single-service approach would avoid the capability compromises inherent in a joint program, as discussed above.

[13] To check the robustness of our comparisons, we repeated this estimation process for production costs using a different methodology to estimate the costs of our notional single-service fighter procurement programs. For this alternative methodology, we used F-22 program data and historical aircraft program cost data based on weight-cost relationship curves created by RAND, a standard cost-estimating methodology. Using this alternative methodology, we arrived at very similar cost relationships when comparing the normalized 2010 SAR LCC estimate for JSF with our three notional single-service fighters. The details of this second methodology and its results are documented in a separate technical report, Lorell, Kennedy, et al., 2013, Appendix F.

**Figure 3.4**
**Estimated Nine Years Past Milestone B, Life Cycle Cost for the Joint Strike Fighter Would Be Higher Than Those of Three Notional Single-Service Programs (Assuming F-22 Cost-Growth Percentages)**

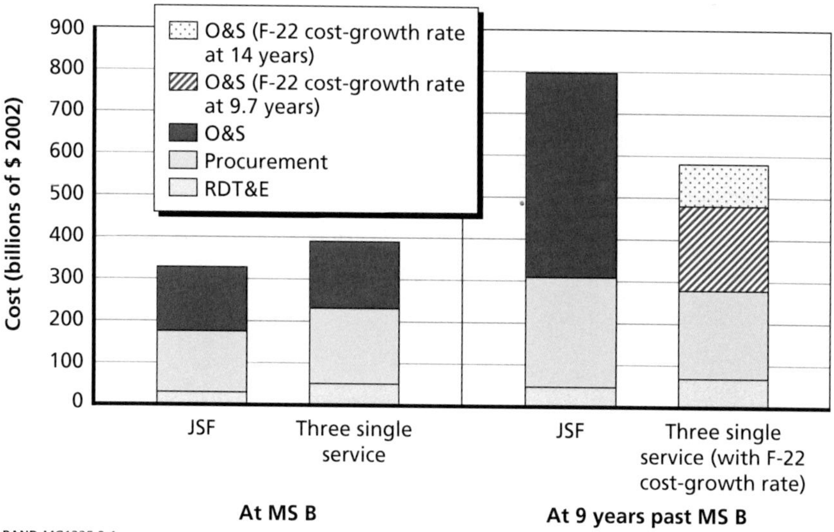


RAND *MG1225-3.4*

## Factors Contributing to Cost Growth

Our analysis of historical joint fighter and other joint aircraft programs shows that cost savings are extremely difficult to achieve when individual service requirements diverge. Greater technical and programmatic complexity and a lack of commonality among service variants contribute to a joint acquisition cost-growth premium, which outweighs the potential savings that theoretically result from joint investment and economies of scale. This is true for the JSF program as well.

JSF resulted from the melding of three separate single-service programs: the Air Force MRF, the Navy A/F-X program, and the Marine Corps ASTOVL program.[14] The requirements for each program were quite different. The Air Force wanted a lightweight, low-cost strike fighter to replace the F-16 and the A-10 and to complement the F-22

[14] Also briefly called the Supersonic STOVL Fighter (SSF) program.

(at this point, the Air Force still hoped to acquire larger numbers of F-22s as the premier air-superiority fighter in the inventory to replace its high-end F-15 fighter). The Air Force did not intend for the MRF to have greater maneuverability and other combat-relevant airframe performance capabilities than existing F-16s, although it did expect that the aircraft would have greater LO capabilities.

By contrast, the Navy sought a larger, heavier, more capable, and complex fighter with its original A/F-X program. This program emerged out of the cancellation of the ATA program in 1989. The Navy wanted a two-seat, two-engine, all-weather, long-range, LO strike fighter to replace the high-end, carrier-based A-6 strike fighter, although it is not clear whether or not the Navy sought sustained supersonic dash capability, because both the ATA and the A-6 were subsonic. Consequently, the Navy came to the JSF program with quite different requirements from those of the Air Force.

Finally, the Marine Corps wanted a longer-range, supersonic STOVL support fighter to replace its subsonic AV-8B Harrier. The Marine STOVL fighter is unique and technologically challenging in many respects. It is the first attempt to develop an operational supersonic STOVL fighter that has progressed beyond the initial developmental stages into production, and the first attempt to develop an LO STOVL fighter. It is also the first fighter ever to use the shaft-driven lift fan propulsion technology that is applied to the STOVL variant of JSF.[15]

The difficulty of reconciling these disparate performance and technological requirements is illustrated by the decline of structural commonality by weight over the course of JSF's initial development. The initial goal of JSF was to achieve 80-percent commonality among service variants. However, as shown in Figure 3.5, airframe commonality by structural weight has declined over time.[16] At MS B, the three

---

15 The shaft-driven lift fan approach was first conceived by the Lockheed Skunk Works in the early 1990s. It was a novel solution to many technical issues posed by developing a supersonic STOVL fighter. However, it had never been used on any operational aircraft prior to JSF. See Paul M. Bevilaqua, "Genesis of the F-35 Joint Strike Fighter," *Journal of Aircraft*, Vol. 46, No. 6, November–December 2009, pp. 1825–1836.

16 Our data cover only the period from MS B through July 2008.

**Figure 3.5**
**Joint Strike Fighter Structural Commonality Has Declined Since Milestone B**

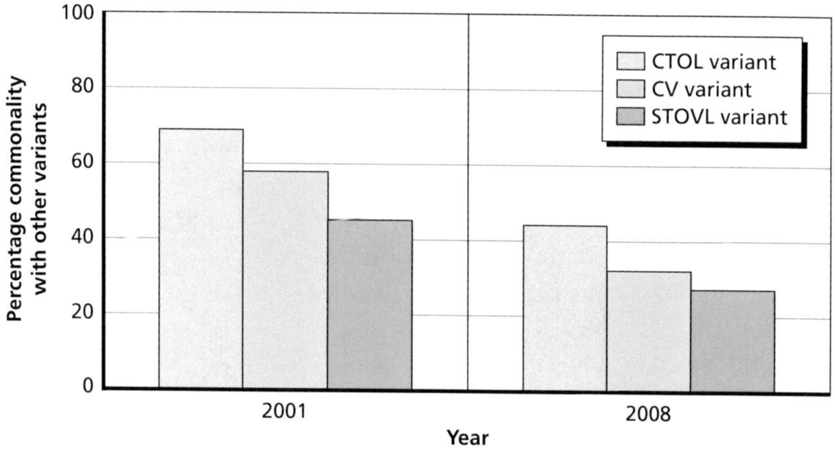


NOTE: We use the CAPE definition of commonality of cousin parts for cost-estimation purposes. Commonality here is measured by airframe structural weight.
RAND *MG1225-3.5*

variants (the CV [aircraft carrier capable], STOVL, and CTOL) ranged from between about 45 to 70 percent commonality by airframe weight. Weight growth on the STOVL variant and various design changes have reduced airframe commonality to a range of about 27 to 43 percent by July 2008.[17] As of this writing, it is not clear how common the mission systems, avionics, software, and engine will be among the three service variants because many of these items are still under development or continue to evolve.[18] Our analysis suggests that increased technological

---

[17] We note that there has been a substantial debate among cost analysts regarding how much credit (in a cost analysis sense) should be applied to "cousin" parts, which are similar but not identical across aircraft. The JSF JPO and contractor regard cousin parts as 82 percent common. By contrast, OSD's CAPE office regards cousin parts as only 25 percent common. Figure 3.5 is based on CAPE's stricter definition of commonality, which is viewed by many senior officials in the Air Force as more authoritative. A similar chart based on the JPO's definition would show greater commonality but a similar pattern of decline.

[18] Although the original goal for avionics, software, and the engine core, as well as many subsystems, was nearly 100-percent commonality, there are indications that commonality among these items for different variants may decline as development progresses.

and programmatic complexity and declining commonality are significant drivers of the RDT&E and procurement cost growth described in this report.

It is not clear why the JSF O&S cost estimate has experienced such high cost growth. In the second half of 2011, OSD undertook a major study of the O&S cost drivers and how to mitigate them. At the time of the completion of our research, the final report from that study had not yet been published. There are several probable reasons that O&S cost growth has been so high, many of which are not necessarily associated with jointness.[19] A definitive answer for the causes of JSF O&S cost growth must await more-detailed analysis. Without doubt, however, the latest estimate of O&S cost growth is greater than that of the F-22 or any other recent single-service aircraft, at the equivalent time after MS B.[20]

The tension between maintaining maximum commonality and meeting different service requirements has been difficult to resolve in the JSF program, resulting in less common variants, greater technical complexity and risk, less optimization for service needs, and difficulty in achieving the promised savings in LCC. Indeed, our analysis indicates that JSF may cost the services the same or more in total LCC than if they had pursued separate, single-service programs, which might have produced differing designs better optimized to meet their unique individual service operating environments and requirements.

---

[19] This, of course, is also true of JSF acquisition program cost growth. We do not claim that all or even necessarily most of JSF acquisition cost growth is due to issues related to jointness. On the other hand, we believe that the evidence indicates that jointness is an important factor in the higher cost growth experienced by JSF than for historical single-service fighters.

[20] F-22 O&S cost estimates have escalated substantially since IOC. However, unlike F-35 cost estimates, F-22 cost estimates are now based on real operational experience and considerable hard data derived from extensive operational use. Historically, once real data become available in significant quantities, O&S cost estimates tend to escalate. Thus, it would not be surprising if JSF O&S cost estimates experience additional cost growth as the aircraft completes development and enters into full operational service.

CHAPTER FOUR

# Additional Implications of Joint Aircraft Programs

Thus far, we have discussed the cost implications of taking a joint approach to fighter and other military aircraft acquisition. Several additional issues bear consideration for future acquisition planning. In this chapter, we briefly review two of the most important: the implications of joint aircraft programs for the industrial base and for operational and strategic risk.

## Implications for the Industrial Base

The aerospace defense industrial base has undergone a major consolidation in recent decades. Although many factors have affected the structure of the industrial base, the pursuit of joint fighter programs has clearly been associated with the decline in the number of credible fighter aircraft prime contractors, both in the 1960s and the 1990s.

For example, in the 1960s, the cancellation of separate single-service Navy and Air Force programs in favor of the joint TFX/F-111 program adversely affected Douglas Aircraft and Republic Aircraft, the contractors for the Navy F6D Missileer fleet air defense fighter and Air Force F-105 fighter, respectively.[1] Shortly thereafter, Douglas exited

[1] The Douglas F6D Missileer was only in the very earliest stages of the design proposal process when Secretary of Defense Robert McNamara canceled it in favor of the TFX. The Republic F-105 was already well advanced into series production and in heavy operational use when Secretary McNamara canceled further planned production in favor of pursuing TFX and other joint programs with the Navy.

the military fighter aircraft market and was bought by the McDonnell Corporation, and Republic was acquired by the Fairchild Corporation. Although the Fairchild Republic division continued on for some years, it was so weakened by the F-105 cancellation that it also eventually disappeared as a military aircraft prime contractor.[2]

At the beginning of the JSF program in the early 1990s, there were eight major prime contractors that were credible combat aircraft developers, shown in Figure 4.1. As of 1990, three major programs for new fighters were in the very early stages of development (the Air Force MRF, Navy A/F-X, and Marine Corps ASTOVL), with five to six prime contractors directly involved in developing designs and possible prototypes.[3] Had those programs continued, between three and six prime contractors would have been developing three entirely separate new fighters, thus maintaining skills and fostering competition and innovation. Instead, as the F-22 and the F/A-18E/F matured and went into production, JSF became the only remaining new fighter/attack aircraft program for any of the services for the foreseeable future. Although three prime contractors (McDonnell Douglas, Boeing, and Lockheed Martin) competed for that program, the down-selection process essentially forced McDonnell Douglas to exit the fighter market and led to its acquisition by Boeing. Boeing's failure to win the final down-select for the JSF prime contractor left it without any future manned fighter/attack program after its current F/A-18E/F and F-15 fighter programs wound down (inherited through its acquisition of McDonnell Douglas). Thus, Lockheed Martin is now the only prime contractor actively leading a fifth-generation manned fighter/attack aircraft development and production program for the foreseeable future. Such a situation reduces the potential for future competition, may discourage innova-

---

[2] Republic became a division of Fairchild in 1965. The Fairchild Republic division later won the Air Force A-X competition for a dedicated, subsonic close support aircraft and subsequently developed the A-10, still in service today. However, the less complex and relatively low-cost A-10 program did not decisively reverse Fairchild Republic's long decline. In 1987, Grumman acquired what was left of Fairchild Republic.

[3] Two other very large-scale fighter programs were also under way at this time in more-advanced stages of the acquisition process: the Lockheed/Boeing F-22 and the McDonnell Douglas/Northrop F/A-18E/F.

**Figure 4.1**
**Recent Consolidation of Fighter Primes Coincides with Joint Strike Fighter Development**

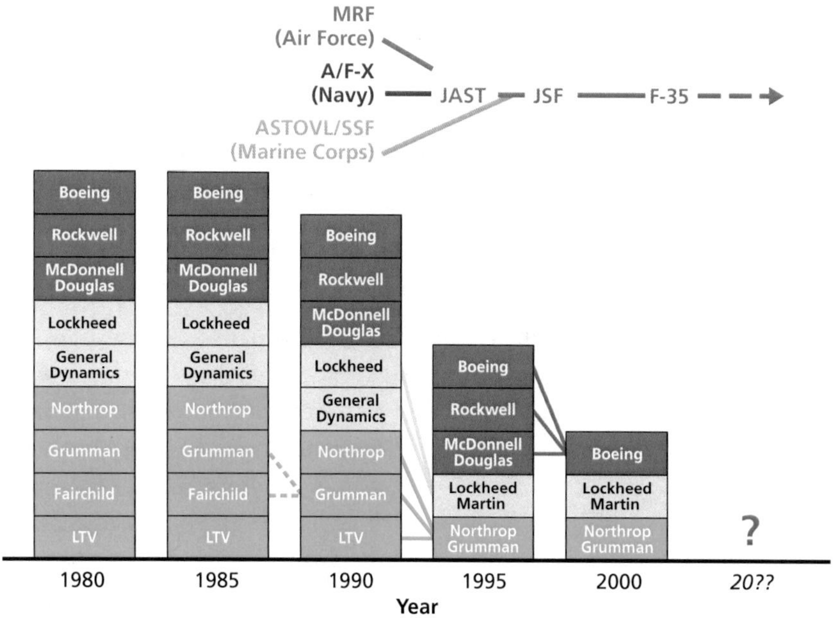


NOTE: LTV = Ling Tempco Vought Corporation. Colors indicate groups of contractors that were eventually consolidated.
RAND *MG1225-4.1*

tion, and makes costs more difficult to control. Whether DoD pursues a joint or single-service acquisition strategy for development of the next fighter aircraft, decisionmakers will have to address the challenges of a smaller, less competitive industrial base and will have to understand the impact of acquisition strategy on the long-term health of the industry.

## Implications for Operational and Strategic Risk

We further examined the operational and strategic risks that are posed by depending on a single weapon-system type or platform across all the services to carry out an entire mission area. Informed by both statistical analysis of fleet-wide stand-downs and analysis of historical crisis situations, we determined that having a diversity of fighter types across the services reduces vulnerability to design flaws and unforeseen enemy capabilities.

First, we developed a simple statistical measure to determine the average probability that an entire fleet of one fighter type would be subjected to a stand-down due to maintenance or safety issues. We examined all four of the current fighter types in the Air Force inventory (F-15, F-16, F-22, and A-10) to determine the number of days since their IOC that each of these fighter types had experienced a fleet-wide stand-down due to maintenance problems or other design or safety considerations.[4] Using these data, we calculated the average probability that each fleet of the same fighter type would experience a fleet-wide stand-down in the future by dividing the total number of days the fighter fleet had been stood down since IOC by the total number of days since IOC.[5] We then calculated that any given modern fighter in the inventory has, on average, a 2.2-percent probability of the entire fleet being ordered to stand down at any given point in the future. If there is only one fighter in the inventory across all the services, then there is a 2.2-percent probability that the entire fleet will be directed to stand down at any given point in time. However, if there are two fighter types in the inventory, the probability declines to 0.05 percent.[6] With the current mix of four distinct fighter types in the Air Force inventory alone, then the probability of all four of those fighters simultaneously being ordered to stand down declines to nearly zero. Although this is a

---

4 We examined a wide variety of Air Force statistical databases to find the most-comprehensive official data for fleet-wide stand-downs and their causes and duration, but we did not find any such existing database. As a result, we depended on open published accounts of major fleet-wide stand-downs to obtain our data. Consequently, our estimate is conservative and understates the probabilities of stand-downs.

5 We use the term *stand-down* rather than the term *grounding*. A stand-down and a grounding are technically different. A stand-down order may be voluntary and precautionary, whereas a grounding is mandatory and usually the result of a catastrophic technical issue. Many of our published sources did not make these distinctions. We therefore use only the term *stand-down* for all events we documented.

6 Arrived at by raising 2.2 percent (0.022) to the power of two, giving 0.000484, or 0.05 percent rounded off. This assumes that the probability of stand-down for each fleet is independent of the other.

very simple approach arithmetically and conceptually, it illustrates how having multiple fighter types radically reduces the probability that all fighters will have to stand down at the same time, reducing operational risk and increasing deterrence.

In addition to the benefit of a lower probability of fleet-wide stand-downs, there are also historical examples of a diversity of fighter types across services that provided warfighters with greater flexibility to respond either to unforeseen problems or design flaws in U.S. forces or to the introduction of unanticipated new enemy capabilities. One example we reviewed was the case of the surprise introduction of the Soviet-developed Mikoyan-Gurevich (MiG)-15 fighter by the North Koreans and Chinese in the early phases of the Korean War. At that time, the U.S. Air Force had four modern jet fighters deployed in its inventory (F-80, F-84, F-86, and F-94) plus several propeller fighters. The Navy also deployed four jet fighter types (F2H, F3D, F7F, F9F) plus several propeller planes. However, with the surprise introduction of the MiG-15 in November 1950, all U.S. fighters in the inventory were found to be at a distinct performance disadvantage. Luckily, superior U.S. fighter pilot training, combined with the ability to rapidly upgrade the F-86 Sabre to performance standards close to those of the MiG-15, permitted U.S. regional commanders to continue their strategy of heavy dependence on air interdiction, a strategy that could not have been pursued had they been required to depend on any of the other existing U.S. fighters. Thus, having a variety of designs and capabilities in the fighter inventory provides a range of options for U.S. warfighters and industry to rapidly and successfully respond to an unanticipated and superior enemy capability.

CHAPTER FIVE

# Conclusions

The quantitative analysis summarized in this report shows that, contrary to expectations, historical joint aircraft programs have not saved money compared with single-service aircraft programs. Although a joint approach has the theoretical potential to lower acquisition costs by pooling RDT&E resources and increasing production runs, historical joint aircraft programs show substantially higher acquisition cost growth than single-service programs. This higher cost-growth percentage is about twice the maximum theoretical joint acquisition savings percentage for a typical fighter program. In addition, whereas a joint approach can save O&S costs by taking advantage of economies of scale, analysis of actual fighter O&S data shows that the potential savings percentages are small and are not likely to counteract the higher average joint aircraft RDT&E and procurement cost-growth rates. On net, joint aircraft programs have not saved overall LCC compared with single-service aircraft programs.

An important reason for the joint acquisition cost-growth premium is that joint fighter and other complex military aircraft programs have had difficulty reconciling diverse service-specific missions, operating environments, and performance requirements in a common design. The attempt to accommodate different service requirements escalates technical and programmatic complexity and risk, which can drive up costs and prolong RDT&E. At the same time, service-specific requirements and demands tend to produce less commonality and lead to more variants, thus reducing the main source of joint cost savings in procurement and O&S. Even so, necessary design compromises often leave participating services unsatisfied and have often resulted in the withdrawal of partners from historical joint fighter programs. Indeed, of the 11 major historical joint fighter programs we examined dating from the 1960s to the 1990s, all but three led to the withdrawal of one of the partners or cessation of the program very early in the effort, almost always because of conflicts over performance requirements and design. These factors work against the potential for joint cost savings and can increase the overall cost of the program compared with a single-service approach.

The same factors that have increased costs for historical joint fighter and other complex joint military aircraft programs continue to affect JSF. Our analysis of SAR data and conservative estimates of notional single-service fighter costs indicate that JSF is not on the path to achieving the savings in overall LCC anticipated at MS B.

Finally, we have identified considerations other than cost that acquisition decisionmakers should bear in mind when formulating future acquisition strategy. The first is that, historically, joint programs have been associated with declining numbers of credible fighter/attack aircraft prime contractors, a situation that is likely to reduce competition and innovation in the future. The second is that joint programs could potentially increase the operational and strategic risk to warfighters by increasing the chances of simultaneous fleet-wide stand-downs of all aircraft in a major mission area and reducing the capability options available to meet unforeseen enemy capabilities and other challenges.

Informed by these findings, we recommend that, unless the participating services have identical, stable requirements, DoD avoid future joint fighter and other complex joint aircraft programs.

# Bibliography


Aghion, Philippe, "Empirical Estimates of the Relationship Between Product Market Competition and Innovation," in Jean-Philippe Touffut, ed., *Institutions, Innovation and Growth: Selected Economic Papers*, Cheltenham, UK: Edward Elgar, Saint-Gobain Centre for Economic Studies Series, 2003, pp. 142–169.

Aghion, Philippe, Nick Bloom, Richard Blundell, Rachel Griffith, and Peter Howitt, "Competition and Innovation: An Inverted-U Relationship," *Quarterly Journal of Economics*, Vol. 120, No. 2, May 2005, pp. 701–728.

Ahn, Sanghoon, *Competition, Innovation and Productivity Growth: A Review of Theory and Evidence*, Paris: Organisation for Economic Co-Operation and Development Economics Department Working Paper 317, January 17, 2002.

Anderson, Fred, *Northrop: An Aeronautical History*, Los Angeles, Calif.: Northrop, 1976.

Anderton, David A., *Republic F-105 Thunderchief*, London: Osprey Publishing, 1983.

———, *North American F-100 Super Sabre*, London: Osprey Publishing, 1987.

Angelmar, Reinhard, "Market Structure and Research Intensity in High-Technological-Opportunity Industries," *Journal of Industrial Economics*, Vol. 34, No. 1, September 1985, pp. 69–79.

Arena, Mark V., Obaid Younossi, Kevin Brancato, Irv Blickstein, and Clifford A. Grammich, *Why Has the Cost of Fixed-Wing Aircraft Risen? A Macroscopic Examination of the Trends in U.S. Military Aircraft Costs over the Past Several Decades*, Santa Monica, Calif.: RAND Corporation, MG-696-NAVY/AF, 2008. As of February 29, 2012:
http://www.rand.org/pubs/monographs/MG696.html

Armitage, M. J., *Unmanned Aircraft*, London: Brassey's Defence Publishers, 1988.

Aronstein, David C., and Michael J. Hirschberg, "An Overview of the US/UK ASTOVL Programs, 1983–1994," *International Powered Lift Conference, AHS International*, Paper IPLC-2000-00039, 2000.

Aronstein, David C., Michael J. Hirschberg, and Albert C. Piccirillo, *Advanced Tactical Fighter to F-22 Raptor: Origins of the 21st Century Air Dominance Fighter*, Reston, Va.: American Institute of Aeronautics and Astronautics, 1998.

Art, Robert J., *The TFX Decision: McNamara and the Military*, Boston, Mass.: Little, Brown, 1968.

Baker, David, "From ATF to Lightning II: A Bolt in Anger—Part One: Design Options and the YF-23A," *Air International*, Vol. 47, No. 6, December 1994.

Barkey, H. D., *Evolution of the F-4 Phantom*, presented at the Technical Program Management Seminar of the American Institute of Industrial Engineers, Washington, D.C., April 19–21, 1971.

Bevilaqua, Paul M., "Joint Strike Fighter Dual-Cycle Propulsion System," *Journal of Propulsion and Power*, Vol. 21, No. 5, September–October 2005.

———, "Genesis of the F-35 Joint Strike Fighter," *Journal of Aircraft*, Vol. 46, No. 6, November–December 2009, pp. 1825–1836. As of January 24, 2013: http://arc.aiaa.org/doi/pdf/10.2514/1.42903

Birkler, John, Anthony G. Bower, Jeffrey A. Drezner, Gordon T. Lee, Mark A. Lorell, and Giles K. Smith, *Competition and Innovation in the U.S. Fixed-Wing Military Aircraft Industry*, Santa Monica, Calif.: RAND Corporation, MR-1656-OSD, 2003. As of February 29, 2012:
http://www.rand.org/pubs/monograph_reports/MR1656.html

Birkler, John, Paul Bracken, Gordon T. Lee, Mark A. Lorell, Soumen Saha, and Shane Tierney, *Keeping a Competitive U.S. Military Aircraft Industry Aloft: Findings from an Analysis of the Industrial Base*, Santa Monica, Calif.: RAND Corporation, MG-1133-OSD, 2011. As of February 29, 2012:
http://www.rand.org/pubs/monographs/MG1133.html

Birkler, John, Edmund Dews, and Joseph P. Large, *Issues Associated with Second-Source Procurement Decisions*, Santa Monica, Calif.: RAND Corporation, R-3996-RC, December 1990. As of February 29, 2012:
http://www.rand.org/pubs/reports/R3996.html

Birkler, John, John C. Graser, Mark V. Arena, Cynthia R. Cook, Gordon T. Lee, Mark A. Lorell, Giles K. Smith, Fred Timson, Obaid Younossi, and Jon Grossman, *Assessing Competitive Strategies for the Joint Strike Fighter: Opportunities and Options*, Santa Monica, Calif.: RAND Corporation, MR-1362-OSD/JSF, 2001. As of February 29, 2012:
http://www.rand.org/pubs/monograph_reports/MR1362.html

Bolkcom, Christopher C., *Tactical Aircraft Modernization: Issues for Congress*, Washington, D.C.: Congressional Research Service, Issue Brief 92115, July 3, 2002.

———, *Proposed Termination of Joint Strike Fighter (JSF) F136 Alternate Engine*, Washington, D.C.: Congressional Research Service, RL33390, April 13, 2006.

Bolten, Joseph G., Robert S. Leonard, Mark V. Arena, Obaid Younossi, and Jerry M. Sollinger, *Sources of Weapon System Cost Growth: Analysis of 35 Major Defense Acquisition Programs*, Santa Monica, Calif.: RAND Corporation, MG-670-AF, 2008. As of April 9, 2012: http://www.rand.org/pubs/monographs/MG670.html

Braybrook, Roy, "F-14 and F-15: The New Wave of Warplanes," *Air Enthusiast*, March 1972.

Bright, Charles D., *The Jet Makers: The Aerospace Industry from 1945 to 1972*, Lawrence, Kan.: Regents Press of Kansas, 1978.

Burton, James G., *The Pentagon Wars: Reformers Challenge the Old Guard*, Annapolis, Md.: Naval Institute Press, 1993.

Congressional Budget Office, *Balance and Affordability of the Fighter and Attack Aircraft Fleets of the Department of Defense*, Washington, D.C., April 1992.

Coulam, Robert F., *Illusions of Choice: The F-111 and the Problems of Weapons Acquisition Reform*, New York: Princeton University Press, 1977.

Davis, Maj Gen Charles R., program executive officer, F-35 Program Office, "JSF Production," briefing presented at the Aviation Week Aerospace and Defense Finance Conference, New York, November 2008.

Delusach, Al, "Long Preparation Paid Off for McDonnell on the F-15," *St. Louis Post-Dispatch*, January 11–12, 1970.

Donald, David, ed., *The Complete Encyclopedia of World Aircraft*, New York: Barnes and Noble, 1999.

Dorr, Robert F., *Lockheed F-117 Nighthawk*, London: Aerospace Publishing, 1995.

Drezner, Jeffrey A., Giles K. Smith, Lucille E. Horgan, J. Curt Rogers, and Rachel Schmidt, *Maintaining Future Military Aircraft Design Capability*, Santa Monica, Calif.: RAND Corporation, R-4199-AF, 1992. As of February 29, 2012: http://www.rand.org/pubs/reports/R4199.html

Ethell, Jeffrey L., *F-15 Eagle*, London: Ian Allan, 1981.

Farrell, Joseph, Richard J. Gilbert, and Michael L. Katz, "Market Structure, Organizational Structure, and R&D Diversity," in Richard Arnott, ed., *Economics for an Imperfect World: Essays in Honor of Joseph E. Stiglitz*, Cambridge, Mass.: MIT Press, 2003, pp. 195–220.

Francillon, René, *McDonnell Douglas F-15A/B*, Arlington, Texas: Aerofax, 1984.

———, *McDonnell Douglas Aircraft Since 1920: Volume II*, Annapolis, Md.: Naval Institute Press, 1990.

Gansler, Jacques S., William Lucyshyn, and Michael Arendt, *Competition in Defense Acquisitions*, College Park, Md.: University of Maryland Center for Public Policy and Private Enterprise, School of Public Policy, revised February 2009. As of February 29, 2012:
http://www.dtic.mil/dtic/tr/fulltext/u2/a529406.pdf

Gentry, Jerauld R., *Evolution of the F-16 Multinational Fighter*, Washington, D.C.: Industrial College of the Armed Forces, student research report, 1976.

Geroski, Paul A., "Entry, Innovation and Productivity Growth," *Review of Economics and Statistics*, Vol. 71, No. 4, November 1989, pp. 572–578.

Gertler, Jeremiah, *F-35 Joint Strike Fighter (JSF) Program: Background and Issues for Congress*, Washington, D.C.: Congressional Research Service, RL30563, April 26, 2011.

Gething, Michael J., *F-15 Eagle*, New York: Arco Publishing, 1983.

Gilbert, Richard J., "Competition and Innovation," *Journal of Industrial Organization Education*, Vol. 1, No. 1, 2006.

Godfrey, David W. H., "Dogfighter Supreme: The Tomcat," *Air Enthusiast International*, January 1977.

Goodwin, Jacob, *Brotherhood of Arms: General Dynamics and the Business of Defending America*, New York: Times Books, 1985.

Green, William, and Gordon Swanborough, *The Complete Book of Fighters: An Illustrated Encyclopedia of Every Fighter Aircraft Built and Flown*, New York: Smithmark, 1994.

Gunston, Bill, *Attack Aircraft of the West*, New York: Scribner, 1974.

———, *The Encyclopedia of World Air Power*, New York: Crescent Books, 1980.

———, *F-111*, New York: Arco Publishing, 1983.

———, *Modern Fighting Aircraft*, New York: Military Press, 1984.

———, *F/A-18 Hornet*, London: Ian Allen, 1985.

Hale, Robert F., assistant director, National Security Division, Congressional Budget Office, testimony before the Subcommittee on Conventional Forces and Alliance Defense, Committee on Armed Services, U.S. Senate, Washington, D.C., April 22, 1991. As of April 9, 2012:
http://handle.dtic.mil/100.2/ADA529821

Hallion, Richard P., *The Evolution of Commonality in Fighter and Attack Aircraft Development and Usage*, Edwards Air Force Base, Calif.: Air Force Flight Test Center History Office, November 1985.

———, "A Troubling Past: Air Force Fighter Acquisition Since 1945," *Air Power Journal*, Winter 1990, pp. 4–23. As of February 29, 2012:
http://www.airpower.au.af.mil/airchronicles/apj/apj90/win90/1win90.htm

Harman, Alvin J., *Analysis of Aircraft Development*, Santa Monica, Calif.: RAND Corporation, P-4976, 1973. As of February 29, 2012:
http://www.rand.org/pubs/papers/P4976.html

Harman, Alvin J., and S. Henrichsen, *A Methodology for Cost Factor Comparison and Prediction*, Santa Monica, Calif.: RAND Corporation, RM-6269-ARPA, August 1970. As of February 29, 2012:
http://www.rand.org/pubs/research_memoranda/RM6269.html

Head, Richard G., *Decision Making on the A-7 Attack Aircraft Program*, Syracuse, N.Y.: Syracuse University, Ph.D. thesis, December 18, 1970.

———, "Doctrinal Innovation and the A-7 Attack Aircraft Decision," in Richard G. Head and Ervin J. Rokke, eds., *American Defense Policy*, 3rd ed., Baltimore, Md.: Johns Hopkins University Press, 1973.

Hehs, Eric, "F-22 Design Evolution, Part I," *Code One Magazine*, April 1998a. As of April 9, 2012:
http://www.codeonemagazine.com/article.html?item_id=40

———, "F-22 Design Evolution, Part II," *Code One Magazine*, October 1998b. As of April 9, 2012:
http://www.codeonemagazine.com/article.html?item_id=41

———, "F-16 Evolution," *Code One Magazine*, September 2008. As of April 9, 2012:
http://www.codeonemagazine.com/article.html?item_id=23

Held, Thomas, Bruce Newsome, and Matthew W. Lewis, *Commonality in Military Equipment: A Framework to Improve Acquisition Decisions*, Santa Monica, Calif.: RAND Corporation, MG-719-A, 2008. As of April 13, 2012:
http://www.rand.org/pubs/monographs/MG719.html

Johnson, Leland, *The Century Series Fighters: A Study in Research and Development*, Santa Monica, Calif.: RAND Corporation, RM-2549, May 1960. As of February 29, 2012:
http://www.rand.org/pubs/research_memoranda/RM2549.html

Jones, Lloyd S., *U.S. Fighters*, Fallbrook, Calif.: Aero Publishers, 1975.

———, *U.S. Naval Fighters*, Fallbrook, Calif.: Aero Publishers, 1977.

Kelly, Orr, *Hornet: The Inside Story of the F/A-18*, Shrewsbury, UK: Airlife, 1991.

Klein, Burton H., William Meckling, and Emmanuel G. Mesthene, *Military Research and Development Policies*, Santa Monica, Calif.: RAND Corporation, R-333, December 1958. As of February 29, 2012:
http://www.rand.org/pubs/reports/R333.html

Knaack, Marcelle Size, *Post–World War II Fighters: 1945–1973*, Washington, D.C.: Office of Air Force History, 1978.

———, *Post–World War II Bombers: 1945–1973*, Washington, D.C.: Office of Air Force History, 1988.

Li, Allen, director, acquisition and sourcing management, U.S. General Accounting Office, *Tactical Aircraft: Status of the F/A-22 and Joint Strike Fighter Programs*, testimony before the Subcommittee on Tactical Air and Land Forces, Committee on Armed Services, U.S. House of Representatives, Washington, D.C.: U.S. General Accounting Office, GAO-04-597T, March 24, 2004. As of February 29, 2012:
http://purl.access.gpo.gov/GPO/LPS48489

Lorell, Mark A., *Bomber R&D Since 1945: The Role of Experience*, Santa Monica, Calif.: RAND Corporation, MR-670-AF, 1995. As of February 29, 2012:
http://www.rand.org/pubs/monograph_reports/MR670.html

———, *The U.S. Combat Aircraft Industry, 1909–2000: Structure, Competition, Innovation*, Santa Monica, Calif.: RAND Corporation, MR-1696-OSD, 2003. As of February 29, 2012:
http://www.rand.org/pubs/monograph_reports/MR1696.html

Lorell, Mark A., and Donna K. Hoffman, *The Use of Prototypes in Selected Foreign Fighter Aircraft Development Programs: Rafale, EAP, Lavi, and Gripen*, Santa Monica, Calif.: RAND Corporation, R-3687-P&L, September 1989. As of February 29, 2012:
http://www.rand.org/pubs/reports/R3687.html

Lorell, Mark A., Michael Kennedy, Robert S. Leonard, Ken Munson, Shmuel Abramzon, David L. An, and Robert A. Guffey, *Do Joint Fighter Programs Save Money? Technical Appendixes on Methodology*, Santa Monica, Calif.: RAND Corporation, MG-1225/1-AF, 2013. As of late 2013:
http://www.rand.org/pubs/monographs/MG1225z1.html

Lorell, Mark A., and Hugh P. Levaux, *The Cutting Edge: A Half Century of U.S. Fighter Aircraft R&D*, Santa Monica, Calif.: RAND Corporation, MR-939-AF, 1998. As of February 29, 2012:
http://www.rand.org/pubs/monograph_reports/MR939.html

Lorell, Mark A., Julia F. Lowell, Michael Kennedy, and Hugh P. Levaux, *Cheaper, Faster, Better? Commercial Approaches to Weapons Acquisition*, Santa Monica, Calif.: RAND Corporation, MR-1147-AF, 2000. As of February 29, 2012:
http://www.rand.org/pubs/monograph_reports/MR1147.html

Lynch, David J., "How the Skunk Works Fielded Stealth," *Air Force Magazine*, Vol. 75, No. 11, November 1992. As of April 9, 2012:
http://www.airforce-magazine.com/MagazineArchive/Pages/1992/November%20 1992/1192stealth.aspx

Martin, Thomas, and Rachel Schmidt, *A Case Study of the F-20 Tigershark*, Santa Monica, Calif.: RAND Corporation, P-7495-RGS, June 1987. As of February 29, 2012:
http://www.rand.org/pubs/papers/P7495.html

Mason, Francis K., *Phantom: A Legend in Its Own Time*, Osceola, Wis.: Motorbooks International, 1984.

Mayer, Kenneth R., *The Political Economy of Defense Contracting*, New Haven, Conn.: Yale University Press, 1991.

Mendenhall, Charles A., *Delta Wings: Convair's High-Speed Planes of the Fifties and Sixties*, Osceola, Wis.: Motorbooks International, 1983.

Miller, Jay, *General Dynamics F-16 Fighting Falcon*, Arlington, Texas: Aerofax, 1982.

———, *Lockheed SR-71 (A/12YF-12/D-21)*, Arlington, Texas: Aerofax, 1983.

Morrocco, John D., "Northrop Sees ASTOVL as Inroad to JAST," *Aviation Week and Space Technology*, Vol. 141, No. 5, August 1, 1994, p. 27.

Neufeld, Jacob, "The F-15 Eagle: Origins and Development, 1964–1972," *Air Power History*, Vol. 48, No. 1, Spring 2001, pp. 4–21.

Newsome, Bruce, Matthew W. Lewis, and Thomas Held, *Speaking with a Commonality Language: A Lexicon for System and Component Development*, Santa Monica, Calif.: RAND Corporation, TR-481-A, 2007. As of April 13, 2012:
http://www.rand.org/pubs/technical_reports/TR481.html

Office of the Under Secretary of Defense for Acquisition, *1983 Defense Science Board Summer Study Briefing Report for Joint Service Acquisition Programs*, Washington, D.C., August 1–12, 1983. As of April 9, 2012:
http://www.dtic.mil/dtic/tr/fulltext/u2/a199739.pdf

Pace, Steve, *Lockheed Skunk Works*, Osceola, Wis.: Motorbooks International, 1992.

Parsons, Iain, *The Encyclopedia of the World's Combat Aircraft*, New York: Chartwell, 1978.

Pattillo, Donald M., *Pushing the Envelope: The American Aircraft Industry*, Ann Arbor, Mich.: University of Michigan Press, 2000.

Perry, Robert L., *Antecedents of the X-1*, Santa Monica, Calif.: RAND Corporation, P-3154, June 1965. As of February 29, 2012:
http://www.rand.org/pubs/papers/P3154.html

———, *A Prototype Strategy for Aircraft Development*, Santa Monica, Calif.: RAND Corporation, July 1972, not available to the general public.

Perry, Robert L., D. DiSalvo, George R. Hall, Alvin J. Harman, G. S. Levenson, Giles K. Smith, and James P. Stucker, *System Acquisition Experience*, Santa Monica, Calif.: RAND Corporation, RM-6072-PR, November 1969. As of February 29, 2012:
http://www.rand.org/pubs/research_memoranda/RM6072.html

Pierrot, Lane, *Options for Fighter and Attack Aircraft: Costs and Capabilities*, Washington, D.C.: Congressional Budget Office staff memorandum, May 1993.

Pierrot, Lane, and Jo Ann Vines, *A Look at Tomorrow's Tactical Air Forces*, Washington, D.C.: Congress of the United States, Congressional Budget Office study, January 1997.

Pint, Ellen M., and Rachel Schmidt, *Financial Condition of U.S. Military Aircraft Prime Contractors*, Santa Monica, Calif.: RAND Corporation, MR-372-AF, 1994. As of February 29, 2012:
http://www.rand.org/pubs/monograph_reports/MR372.html

Rich, Ben R., and Leo Janos, *Skunk Works: A Personal Memoir of My Years at Lockheed*, Boston, Mass.: Little, Brown, 1994.

Rich, Michael D., Edmund Dews, and C. L. Batten, *Improving the Military Acquisition Process: Lessons from RAND Research*, Santa Monica, Calif.: RAND Corporation, R-3373-AF/RC, 1986. As of February 29, 2012:
http://www.rand.org/pubs/reports/R3373.html

Scherer, F. M., "Market Structure and the Employment of Scientists and Engineers," *American Economic Review*, Vol. 57, No. 3, June 1967, pp. 524–531.

Schmutzler, Armin, *The Relation Between Competition and Innovation: Why Is It Such a Mess?* London: Centre for Economic Policy Research, Discussion Paper 7640, January 2010.

Scutts, Jerry, *F-105 Thunderchief*, London: Ian Allen, 1981.

Shenon, Philip, "Jet Makers Preparing Bids for a Rich Pentagon Prize," *New York Times*, March 12, 1996. As of April 9, 2012:
http://www.nytimes.com/1996/03/12/us/jet-makers-preparing-bids-for-a-rich-pentagon-prize.html

Smith, Giles K., A. A. Barbour, Thomas L. McNaugher, Michael D. Rich, and William Stanley, *The Use of Prototypes in Weapon System Development*, Santa Monica, Calif.: RAND Corporation, R-2345-AF, March 1981. As of February 29, 2012:
http://www.rand.org/pubs/reports/R2345.html

Smith, Giles K., and E. T. Friedmann, *An Analysis of Weapon System Acquisition Intervals, Past and Present*, Santa Monica, Calif.: RAND Corporation, R-2605-DR&E/AF, November 1980. As of February 29, 2012:
http://www.rand.org/pubs/reports/R2605.html

Sponsler, George C., N. Rubin, Dominique Gignoux, and E. Dare, *The F-4 and the F-14*, Gaithersburg, Md.: Columbia Research Corporation, May 1973.

Stevenson, James Perry, *Grumman F-14 "Tomcat,"* Fallbrook, Calif.: Aero Publishers, 1975.

———, *McDonnell Douglas F-15 Eagle*, Fallbrook, Calif.: Aero Publishers, 1978.

———, *The Pentagon Paradox: The Development of the F-18 Hornet*, Annapolis, Md.: Naval Institute Press, 1993.

———, *The $5 Billion Misunderstanding: The Collapse of the Navy's A-12 Stealth Bomber Program*, Annapolis, Md.: Naval Institute Press, 2001.

Stoff, Joshua, *The Thunder Factory: An Illustrated History of the Republic Aviation Corporation*, Osceola, Wis.: Motorbooks International, 1990.

Stuart, William G., *Northrop F-5 Case Study in Aircraft Design*, Hawthorne, Calif.: Northrop Corporation, September 1978.

Stubbing, Richard A., with Richard A. Mendel, *The Defense Game: An Insider Explores the Astonishing Realities of America's Defense Establishment*, New York: Harper and Row, 1986.

Sullivan, Michael, director, acquisition and sourcing management issues, U.S. Government Accountability Office, *Tactical Aircraft: F/A-22 and JSF Acquisition Plans and Implications for Tactical Aircraft Modernization*, testimony before the Subcommittee on AirLand, Committee on Armed Services, U.S. Senate, Washington, D.C., GAO-05-519T, April 6, 2005. As of February 29, 2012: http://purl.access.gpo.gov/GPO/LPS60224

———, director, acquisition and sourcing management, *Joint Strike Fighter: Impact of Recent Decisions on Program Risks*, testimony before the Subcommittee on Air and Land Forces, and Seapower and Expeditionary Forces, Committee on Armed Services, U.S. House of Representatives, Washington, D.C., GAO-08-569T, March 11, 2008. As of February 29, 2012: http://www.gao.gov/new.items/d08569t.pdf

———, director, acquisition and sourcing management, *Joint Strike Fighter: Strong Risk Management Essential as Program Enters Most Challenging Phase*, testimony before the Subcommittee on Air and Land Forces, Committee on Armed Services, U.S. House of Representatives, Washington, D.C., GAO-09-711T, May 20, 2009. As of February 29, 2012: http://www.gao.gov/new.items/d09711t.pdf

———, director, acquisition and sourcing management, *Joint Strike Fighter: Significant Challenges and Decisions Ahead*, testimony before the Subcommittee on Air and Land Forces and Seapower and Expeditionary Forces, Committee on Armed Services, U.S. House of Representatives, Washington, D.C., GAO-10-478T, March 24, 2010. As of February 29, 2012: http://purl.access.gpo.gov/GPO/LPS122323

———, director, acquisition and sourcing management, *Joint Strike Fighter: Restructuring Places Program on Firmer Footing, but Progress Is Still Lagging*, testimony before the Committee on Armed Services, U.S. Senate, Washington, D.C., GAO-11-677T, May 19, 2011. As of February 29, 2012: http://www.gao.gov/new.items/d11677t.pdf

Swanborough, Gordon, *United States Military Aircraft Since 1909*, London: Putnam, 1963.

———, *United States Navy Aircraft Since 1911*, London: Putnam, 1968.

Sweetman, Bill, *A-10 Thunderbolt II*, New York: Arco Publishing, 1984a.

———, *Phantom*, London: Jane's Publishing Company Limited, 1984b.

———, "Lockheed YF-22: Stealth with Agility," *World Airpower Journal*, Vol. 6, Summer 1991a.

———, "The Fighter They Didn't Want," *World Airpower Journal*, Vol. 7, Autumn–Winter 1991b.

Sweetman, Bill, and James C. Goodall, *Lockheed F-117A: Operation and Development of the Stealth Fighter*, Osceola, Wis.: Motorbooks International, 1990.

Thornborough, Anthony M., and Peter E. Davies, *Grumman A-6 Intruder Prowler*, London: Ian Allen, 1987.

U.S. Department of Defense, *Operation of the Defense Acquisition System*, Instruction 5000.02, December 8, 2008. As of April 9, 2012: http://www.dtic.mil/whs/directives/corres/pdf/500002p.pdf

U.S. Government Accountability Office, *Tactical Aircraft: Opportunity to Reduce Risks in the Joint Strike Fighter Program with Different Acquisition Strategy*, Washington, D.C., GAO-05-271, March 2005. As of February 29, 2012: http://purl.access.gpo.gov/GPO/LPS59279

———, *Joint Strike Fighter: DoD Plans to Enter Production Before Testing Demonstrates Acceptable Performance*, Washington, D.C., GAO-06-356, March 2006. As of February 29, 2012: http://purl.access.gpo.gov/GPO/LPS68153

———, *Joint Strike Fighter: Progress Made and Challenges Remain*, Washington, D.C., GAO-07-360, March 2007. As of February 29, 2012: http://purl.access.gpo.gov/GPO/LPS80611

———, *Joint Strike Fighter: Recent Decisions by DoD Add to Program Risks*, Washington, D.C., GAO-08-388, March 2008. As of February 29, 2012: http://www.gao.gov/new.items/d08388.pdf

———, *Joint Strike Fighter: Additional Costs and Delays Risk Not Meeting Warfighter Requirements on Time*, Washington, D.C., GAO-10-382, March 2010. As of February 29, 2012: http://www.gao.gov/new.items/d10382.pdf

U.S. House of Representatives, *The Navy's A-12 Aircraft Program: Joint Hearing Before the Procurement and Military Nuclear Systems Subcommittee and the Research and Development Subcommittee and the Investigations Subcommittee of the Committee on Armed Services, U.S. House of Representatives*, Washington, D.C., December 10, 1990.

———, *A-12 Acquisition: Hearings Before the Investigations Subcommittee of the Committee on Armed Services, U.S. House of Representatives*, Washington, D.C., April 9, April 18, July 18, July 23, July 24, 1991a.

———, *Oversight Hearing on the A-12 Navy Aircraft: Hearings Before the Legislation and National Security Subcommittee of the Committee on Government Operations, U.S. House of Representatives*, Washington, D.C., April 11, July 24, 1991b.

*Variable Sweep: A Case History of Multiple Re-Innovation*, Santa Monica, Calif.: RAND Corporation, P-3459, October 1966. As of February 29, 2012: http://www.rand.org/pubs/papers/P3459.html

Yenne, Bill, *Lockheed*, New York: Crescent Books, 1980.

———, *McDonnell Douglas: A Tale of Two Giants*, New York: Crescent Books, 1985.